Mathematical Programming

Edited by

T. C. Hu and
Stephen M. Robinson

Proceedings of an Advanced Seminar
Conducted by the Mathematics Research Center
The University of Wisconsin, and the
U. S. Army at Madison, September 11–13, 1972

Academic Press
New York · London 1973

A Subsidiary of Harcourt Brace Jovanovich, Publishers

ACADEMIC PRESS, INC.
111 Fifth Avenue, New York, New York 10003

United Kingdom Edition published by
ACADEMIC PRESS, INC. (LONDON) LTD.
24/28 Oval Road, London NW1

Library of Congress Cataloging in Publication Data

Advanced Seminar on Mathematical Programming, Madison, Wis., 1972.
Mathematical programming.

(Publication of the Mathematics Research Center, the University of Wisconsin, no. 30)
"Conducted by the Mathematics Research Center, the University of Wisconsin, and the U. S. Army."
1. Programming (Mathematics)–Congresses.
I. Hu, Te Chiang, DATE ed. II. Robinson, Stephen M., ed. III. United States. Army. Mathematics Research Center, Madison, Wis. IV. Title. V. Series: United States. Army. Mathematics Research Center, Madison, Wis. Publications no. 30.
QA3.U45 no. 30 [T57.7] 510'.8s [658.4'033] 72-12204
ISBN 0–12–358350–0

PRINTED IN THE UNITED STATES OF AMERICA

Contents

MATHEMATICAL PROGRAMMING

Publication No. 30
of the Mathematics Research Center
The University of Wisconsin

ACADEMIC PRESS RAPID MANUSCRIPT REPRODUCTION

Preface

This book contains the proceedings of an Advanced Seminar on Mathematical Programming held in Madison, Wisconsin, September 11-13, 1972, under the auspices of the Mathematics Research Center, University of Wisconsin–Madison, and with financial support from the United States Army under Contract No. DA-31-124-ARO-D-462. Approximately two hundred persons from government, industry, and various universities attended the three-day conference.

The aim of the seminar was to offer insight into branches of mathematical programming which have advanced rapidly in recent years. Of the ten lectures given, four were on integer programming, two on game theory, and one each on large-scale systems, nonlinear programming, dynamic programming, and combinatorial equivalence. A lecture on graph theory by Professor D. R. Fulkerson was originally planned, and its text is published here. Due to his sudden illness, however, a substitute lecture on combinatorial equivalence was delivered by Professor R. M. Karp. As the text of Professor Karp's talk has appeared in *Complexity of Computer Computation* (Plenum Press, 1972), it is not included in this volume. Also, because of the necessity to meet the publication schedule the text of Dr. Philip Wolfe's lecture has not been included.

We are indebted to Professors M. L. Balinski, R. M. Karp, W. F. Lucas, A. W. Tucker, and R. D. Young for their service as session chairmen, as well as to those of our colleagues in the mathematical programming community who contributed to the improvement of this book by refereeing the manuscripts. We should also like to express our sincere thanks to Mrs. Gladys Moran, the meeting secretary, who devoted many hours of her time to the preparation and conduct of the Seminar, and to Mrs. Dorothy Bowar who typed the manuscripts for publication.

We hope that this book will benefit those who could not attend the meeting, and that it will also serve as a permanent source of reference in mathematical programming.

T. C. H.
S. M. R.

MATHEMATICAL PROGRAMMING

On the Need for a System Optimization Laboratory

GEORGE B. DANTZIG, R. W. COTTLE, B. C. EAVES, F. S. HILLIER, A. S. MANNE, G. H. GOLUB, D. J. WILDE, AND R. B. WILSON[1]

Need.

From its very inception, it was envisioned that linear programming would be applied to very large, detailed models of economic and logistical systems [Wood and Dantzig (1947)]. Kantorovich's 1939 proposals, which were before the advent of the electronic computer, mentioned such possibilities. In the intervening 25 or so years, electronic computers have become increasingly more powerful, permitting general techniques for solving linear programs to be applied to larger and larger practical problems. In the author's opinion, however, additional steps are necessary if there is to be significant progress in solving certain pressing problems that face the world today.

The conference on Large-Scale Resource Allocation Problems held at Elsinore, Denmark, July 5-9, 1971 represents an historic first because it demonstrates that optimization of very large-scale planning problems can be achieved on significant problems.[2] I cite some examples from the conference:

Arthur Geoffrion's paper "Optimal Distribution System Design" is of interest because (1) it described the successful solution of a large-scale problem from commerce, (2) it involved discrete variables (representing the integer number of warehouses to be built or closed down), (3) it successfully

[1, 2] See footnotes following references.

combined a variety of advanced techniques in a single computer program.

Leon Lasdon's paper "Uses of Generalized Upper Bounding Methods in Production Scheduling" is of interest because (1) it not only described a successful large-scale application (this time to a rubber factory), (2) made use of advanced techniques, but also, (3) because it showed it was possible to automatically schedule day-to-day operations consistent with the long-term goals, i.e., it successfully combined short and long-term planning goals of an enterprise.

The papers by several authors (for example those of Abadie, Buzby, Huard) are particularly noteworthy because they described the successful solution of real problems (Electric Energy Production and Olefin Production) that were essentially non-linear and large-scale in nature.

Society would benefit greatly if certain total systems can be modeled and successfully solved. For example, crude economic planning models of many developing countries indicate a potential growth rate of GNP of 10% to 15% per year. To implement such a growth (aside from political difficulties) requires a carefully worked out detailed model and the availability of computer programs that can solve the resulting large-scale systems. The world is currently faced with difficult problems related to population growth, availability of natural resources, ecological evaluation and control, urban redesign, design of large-scale engineering systems (e.g., atomic energy, and recycling systems), and the modeling of man's physiological system for the purpose of diagnosis and treatment. These problems are complex, are urgent and can only be solved if viewed as total systems. If not, then only patchwork, piecemeal solutions will be developed (as it has been in the past) and the world will continue to be plagued by one crisis after another caused by poor planning techniques. For solutions, these problems require total system planning, modeling and optimization.

It is my belief that it is necessary at this time to create several system optimization laboratories where enough critical mass would exist that representative large-scale models (of the type referred to above) could be practically modeled and numerically solved. Solving large-scale systems cannot be approached piecemeal or by publishing a few theoretical papers. It is a complex art requiring the development of a whole arsenal of special tools.

Background.

The optimization of large-scale systems is technically an extremely difficult subject. Historically, starting with U. S. Air Force problems in 1947, linear programs were formulated to solve just such systems. These problems involved systems of interlocking relations involving many planning periods, combat units, types of personnel and supply. It led to thousands of equations in many thousands of unknowns. This was beyond computational capabilities. It was necessary to severely restrict the class of practical problems to be solved. Starting around 1954 a series of purely theoretical papers began to appear on how to efficiently solve large systems and by 1970 they numbered about 200. There was little in the way of implementation. Exceptions were the out-of-kilter algorithms for network flow problems proposed by Ford and Fulkerson [1958] and the "decomposition principle" of Philip Wolfe and myself which had been tried but with variable results [1960]. On the other hand a more modest proposal of Richard Van Slyke and myself (generalized-upper bounds) has been very successful [1967]. Apparently a great deal in the way of empirical testing of ideas is necessary and this has not been easy to do because the test models have to be complex to be pertinent and cost a great deal of money to program and solve. Therefore progress has been slow up to the time of the Elsinore meeting.

Since its origins in the development of transport allocation methods in the early 1940's, and especially since the introduction of the Simplex Method of linear programming in 1947, the power of the methods of mathematical programming, and the range of effectiveness of its applications, have grown enormously. In the intervening decades the methodology has been extended to include non-linear and integer programming, dynamic programming and optimal control, and a host of other types of optimization problems.

The range of applications has been extended from simple allocation problems to an enormous variety of problems in intertemporal allocation and investment planning, engineering design and optimization, and scientific studies of physical, biological, and ecological systems. There is, in fact, no end foreseeable to the applications of mathematical programming to a number of important (and crucial) optimization problems.

Some Examples of Important Applications.

A. INVESTMENT PLANNING (INTERTEMPORAL ALLOCATION): Problems of aggregate economic planning for a (developing) country, present an exploitable special structure that has been studied intensively and has great potential. Related structures occur in problems of dynamic programming and optimal control. Related but more complicated structures arise, for example, in problems of plant location and time-phasing, and in investment planning in general in the firm.

B. DECENTRALIZED ALLOCATION: The origin of the modern methods of decomposition, and still one of the major areas of application, is the class of decentralized allocation problems, in which scarce resources are to be allocated among several otherwise independent enterprises or "divisions". Closely related is the class of problems of two-stage allocation under uncertainty, for which in the linear case it is known that the dual problem is one of decentralized allocation. It is of particular importance to realize that the "divisional subproblems" may themselves be of a special structure (e.g., a transportation problem) which can be exploited.

C. ENGINEERING DESIGN AND OPTIMIZATION: A variety of engineering design and process optimization problems present specially-structured mathematical programs for which the structural features are highly dependent on the process being studied. Problems of this type illustrate the need for a flexible and comprehensive software package from which components can be drawn to build up models of very complex systems.

D. PHYSICAL, BIOLOGICAL, AND ECOLOGICAL SYSTEMS: A number of problems in the physical sciences (e.g., X-ray crystallography) and biological sciences (e.g., models of body processes) present specially-structured mathematical programming problems. An extreme example are models of ecological systems in which the many and varied relationships among the components again require a flexible and comprehensive software package.

E. URBAN PLANNING: Coordinated planning of the many component subsystems (e.g., transport, recreation, education, etc.) of an urban environment presents a complex systems optimization problem for which ordinarily the most powerful and flexible methods are required.

F. LOGISTICS: Coordinated logistical support for any large industrial (e.g., warehousing and transport) or government (military) activity normally presents a system optimization problem of considerable size and complexity, but with exploitable structural features.

G. TRANSPORTATION SYSTEMS: Various problems concerning the design of transportation systems can be formulated as network optimization models of a combinatorial nature. These models typically have very special mathematical-programming structures for which highly efficient algorithms can be devised.

The Functions of a Systems Optimization Laboratory.

The purpose of such a laboratory would be to support the development of computational methods and associated computer routines for numerical analysis and optimization of large-scale systems. The ultimate objective of the development effort would be to provide an integrated set of computer routines for systems optimization which:

I. is freely and publicly available to users of government, science, and industry,

II. is thoroughly practical and widely useful in applications to diverse kinds of large-scale systems optimization problems,

III. embodies the most powerful techniques of mathematical programming and numerical analysis, and

IV. has been thoroughly tested for efficiency and effectiveness on representative systems optimization problems arising in practice.

The development effort of such a laboratory in its initial stages would consist of three basic activities: [1] research in mathematical programming, including particularly the analysis, development, and testing of special computational methods for certain specially structured optimization problems that occur frequently in systems optimization, or as subproblems of larger systems, [2] collection of representative systems optimization problems arising in practice in government, science, and industry, in order both to study their mathematical structure and to use them as test problems for studies of efficiency, and [3] development of an integrated set of computer routines, and an associated macro-language to enable its flexible use, which implements the most powerful of existing methods for systems optimization.

The creation of such a laboratory would be a concerted effort to break a bottleneck which is currently constricting the applications of mathematical programming to many of the most important systems optimization problems. This bottleneck is the lack of an integrated collection of compatible computer routines, preferably organized and callable via a macro-language, which can be employed efficiently and flexibly in a wide diversity of practical applications.

The origins and nature of the bottleneck can be described as follows. The existing methods of mathematical programming exploit either general structure or special structure. Those that exploit general structure take advantage of the fact that in a particular problem, the functional forms involved are linear, or quadratic, or convex, separable, etc.

Methods of this kind ordinarily are limited in their applications by the size and speed of the computing equipment available according to some power (often the third or fourth) of the number of variables and/or constraints. Those that exploit special structure take advantage of further particular features of a problem. For example, in the case of linear programming, which is the most highly developed in this respect, there are methods which exploit the special structures of (1) network problems arising in transport planning, (2) "block-diagonal" problems arising in decentralized allocation problems, (3) "staircase" problems arising in dynamic investment planning, economic growth models, and optimal control, (4) problems amenable to "column generation" arising in production scheduling and elsewhere, (5) general problems with "sparse" matrices etc. Moreover, there is a substantial and powerful theory of how to decompose large and complicated systems into their component sybsystems and from analyses of these components to derive solutions to the original system. Methods that exploit special structure are not limited in the range of their applicability in the way that ordinary general-structure methods are; indeed, with present methods and computing equipment it is practical in certain cases to solve systems with close to a million of variables and constraints. (For example the National Biscuit Company problem solved by Mathematica.)

It is the nature of human activity, and in large part of the physical world as well, that large and complicated endeavors are organized as systems of interrelated parts, and indeed, as systematic hierarchies of interrelated subsystems. Such systems typically exhibit special mathematical structures. These special structures permit numerical analysis and optimization via methods that exploit the special structure, whereas general-structure methods would be infeasible if the problem is of the size normally encountered in practice. The extension of the range of applications of mathematical programming is, therefore, most promising for pressing world problems involving total system optimization discussed earlier since they exhibit special structures.

Nature of the Bottleneck.

The bottleneck, however, is that presently there is not available any collection of decomposition methods and special-structure methods implemented in freely available, efficient, tested, flexible computer routines which can be applied easily, cheaply, and with confidence to practical problems as they arise. The result has been, and will continue (if development work does not proceed), that in each potential application it would be necessary to develop computer routines especially for the project. Because this is so costly in expense or time, it is generally not done and the valuable potential application to the system optimization is foregone.

There are three reasons for this unfortunate state of affairs. One is that in the past researchers on decomposition methods and special-structure methods have not had a viable way of enabling their work to contribute directly to the construction of such a collection of computer routines. Either there was no incentive to complement their research results with practically useful computer routines; or, if they did do it, there was no way that the routines could be documented, tested, and ultimately incorporated into a larger collection of established routines; indeed, there has been almost a complete absence of standard documentation procedures, standard test problems, and standard compatibility requirements for callability, data input, and output. The consequence has been that research, implementation, and applications of systems optimization have been uncoordinated and disconnected, to everyone's detriment.

The second reason is that the incentives to development work have operated at cross purposes with the ultimate goals. As mentioned, in a particular application it is usually too costly or time-consuming to undertake the development of the needed computer routines, or just as likely, the organization faced with the tasks lacks the expert competence among its staff to complete the job successfully. On the other hand, occasional development work has been undertaken by private software firms. Indeed, five or ten years ago one would have had great hopes that this approach would succeed. In fact, however, the incentive to private firms has in nearly every instance been to keep their routines proprietary, expensive to use, and noncallable. For the most part, private firms have responded to the natural incentive to appropriate the public know-how into a privately saleable commodity.

The third reason is that there has not been support for a coordinated development effort, one that assembles expert competence in theory, numerical methods, and computer science, and that ensures the permanence of its work through a thorough program of experimentation, testing, documentation, and enforced compatibility requirements.

A Systems Optimization Laboratory could be carefully designed to overcome these impediments to progress in the field. It could bring together the various kinds of expert competence that are needed, and it could implement the development effort in a coordinated program of research, programming, experimentation, testing, and documentation, with the results to be made freely and widely available for diverse applications in a flexible and easily used form.

The major research activities of a System Optimization Laboratory can be classified broadly as follows: (1) basic research related to optimization theory, (2) development of computational methodology for mathematical programming, including general-structure methods, decomposition methods, and special-structure methods, and (3) construction and evaluation of algorithms.

Software Development.

A major activity of System Optimization Laboratory would be the development of software packages for systems optimization. This development effort could proceed on two different levels. First, a major activity would be the completion of a macro-language for organizing and calling routines in the software package. Mainly this could be an extension of the macro-language Mathematical Programming Language [MPL] under development by the author. The second major activity could be the programming, testing, and documentation of algorithms for decomposition and special structures, including experimentation with alternative algorithms, and testing of algorithms on practical problems. Computer routines would be thoroughly documented, tested on standard problems, and written in a format compatible with and callable by the macro-language.

External Affairs.

Three important activities of the Laboratory fall under this heading. First, members of the technical staff could undertake the collection and study of examples of systems optimization problems arising in government, science, and industry, for use both as test problems and as indicators of the types of systems and specially-structured problems of major importance in practice. Many examples are already known, but further empirical data is considered desirable to ensure the ultimate usefulness of the Laboratory's work. Second, other researchers in the field could be solicited to obtain algorithms, computer routines, and test problems for inclusion in the Laboratory's studies. Also, the Laboratory could disseminate information to potential contributions on the requirements for computer routines to be compatible with the Laboratory's software package. Third, when the Laboratory's software package is reasonably complete, it could undertake to make it available to users--this being, of course, the ultimate purpose of the Laboratory.

Research Projects of a System Optimization Laboratory.

A major goal of the Systems Optimization Laboratory would be to provide standardized computer routines for systems optimization. The types of research activities that would be needed to support this effort are outlined below. Particular areas of research that might be planned for the initial project period will be described first:

A. Decomposition Methods: The chief requirement in the construction of numerical methods for optimizing large systems is that the algorithm exploit the special structure of the system. The body of theory and techniques which addresses this requirement are generally called decomposition methods. The range of decomposition methods is quite diverse, however, since of necessity a particular algorithm must reflect the special structure of the class of problems to which it is applicable.

One preliminary task in the development of decomposition methods would be the construction of an efficient taxonomy for system structures. This task is only partially complete. The major taxonomic features that are well understood can be described briefly as follows. First, there is a large and important class of problems whose special structure permits the design of an efficient algorithm based directly on this structure. Usually, duality and compact representation schemes play a key role in the design of the network problems, problems with upper and lower bound constraints, and a number of nonlinear problems (geometric programming, fractional programming, variable-factor programming, etc.). Often problems with these special structures occur as subproblems in larger systems and it is therefore important to have available efficient, tested, and documented routines for these problems which are easily callable.

A second major class of problems are those which, in the linear case, are characterized by sparce matrices. (Hence the numerical structure is quite general except for the known presence of many zeros.) Compact representation schemes for sparse matrices play the major role in the development of algorithms for these problems [Dantzig (1963)].

A third class of problems are those which are amenable to generation techniques. The major examples from this class are the column generating techniques of Gilmore and Gomory (1961, 1965) for "cutting-stock" and related problems, and the row and column generating techniques of Wilson (1972) for 2-person games in extensive forms, both of which use dynamic programming as the means of generating data explicitly that is otherwise embodied implicitly in the problem formulation. A generating technique of much greater generality is the method of generalized programming in which it is required only that the data be generated from a convex set using duality information from a master coordinating problem.

The generalized programming method of Wolfe (see Chapter 22 in Dantzig (1963)) is actually a generalization of the decomposition method for linear programs with block-angular structures [Dantzig and Wolfe (1962)], which represent a fourth major class of system structures--those which (in either primal or dual form, including multi-stage programming under uncertainty [cf. Dantzig and Madansky (1961) and a variety of other dynamic programming problems] represent a problem of allocating scarce resources to otherwise independent subproblems. Zschau's (1967) primal decomposition algorithm also applies to this class of problems, which are of prime importance in applications.

Wolfe's generalized programming approach is also applicable to a fifth major class of problems which is closely related to the previously mentioned class, namely the class of multi-stage allocation problems represented by dynamic investment problems and optimal control problems. Another example is the linear control problem which can be solved using generalized programming [see Dantzig (1966)].

Both of these last two classes are instance of a general class, which can be called nearly decomposable problems. In this general class one finds a macro-structure which would be perfectly decomposable into independent subproblems except for the presence of a relatively few connections (and therefore interdependencies) among the subproblems. The development of efficient algorithms for nearly decomposable problems is a major area for research and one for which the range of applications is enormous. Its successful conclusion may require the development of general methods for highly connected systems, such as have been recently proposed by Douglass Wilde (unpublished). One form of such a method is presently available in Benders' decomposition method (1962).

Surveys of the major decomposition methods are given by Geoffrion (1970) and Lasdon (1971).

In the area of decomposition methods, the Systems Optimization Laboratory would pursue essentially three research and development activities. First, a major effort could be to program, test and document existing decomposition methods as part of the development of the macro-language MPL [Dantzig et al. (1970)]. This development effort is aimed at creating a useful software package for many of the most important systems optimization problems which arise in practice. Second, a part of the research effort could be devoted to the construction of new algorithms for general nearly-decomposable problems and for tightly connected systems.

The third part of the research program would reflect the important role of structural taxonomy in the development of decomposition methods. In connection with the Laboratory's empirical studies of some of the major systems optimization problems encountered in practice, a structural taxonomy could be developed and comparative studies made of the relative efficiencies of alternative methods of optimizing systems of similar structures. There are, moreover, a number of systems optimization problems of known structure, and of great practical importance, for which an intensive development effort could be devoted to the construction of efficient algorithms. First on this list is the class of "staircase" problems represented by dynamic investment models in economics and business and optimal control problems arising in (among other contexts) ecological models.

In general, the Laboratory's work on decomposition methods would provide a synthesizing focus for its entire spectrum of studies on systems optimization. The primary objective would be to provide an unified body of theory, methods, and computer routines for the efficient and practical numerical analysis of large systems.

B. Mathematical Programming, Matrix Decomposition and Sparse Matrix Techniques: (The comments of this section are due to Gene Golub.) For many algorithms in mathematical programming it is necessary to compute a sequence of matrix decompositions. For example, in the classical simplex algorithms for solving linear programming problems it is necessary to solve two or three systems of linear algebraic equations at each iteration. There are many ways of solving these systems, but a particularly effective numerical algorithm is to use some form of the LU decomposition of a matrix. At each stage of the simplex algorithm the coefficient matrix is changed by one column so that one is concerned with techniques of updating the matrix decomposition in an efficient and stable manner, especially when the data matrix is very sparse.

In general, suppose that a matrix A and some factorization of A are given, e.g., $A = PTQ^T$, where P and Q are orthogonal matrices and T is a triangular matrix. The problem then is to compute the factorization of $A + \sigma \underset{\sim}{u} \underset{\sim}{v}^T$ where $\underset{\sim}{u}$ and $\underset{\sim}{v}$ are given vectors and σ is scalar quantity, or a factorization of A when A is changed by one column.

The three basic considerations in computing the new factorization are the following: (1) The updating should be performed in as few operations as possible. This is especially true when handling large masses of data where continual updating is needed. (2) The numerical procedure should be stable. Some procedures which have been recommended in the literature can easily become numerically unstable. This is especially true for the Cholesky factorization of a matrix when $\sigma = -1$. (3) The updating procedure should preserve sparsity. Quite often the original matrix factorization will be sparse, and it is desirable to preserve the sparsity by possibly rearranging the rows and columns of the original data matrix.

The problem of updating occurs in many other contexts, e.g., statistics and control theory. For this reason, it is especially important to have methods which yield fast, accurate, and sparse factorizations.

Therefore, a study would be made of various factorizations and how they may be used in large scale programming problems, especially when the data matrix is structured. The sparse-matrix techniques are especially useful as an alternative when the decomposition principle is applicable. Furthermore, the matrix-decomposition methods would be most useful when the complementarity methods for solving mathematical programming problems are applicable. Some study has already been made in this direction [Tomlin (1971)].

C. Complementarity Methods: The development of complementarity methods is the major advance in the theory and technique of mathematical programming in recent years. The application of this approach to decomposition and special-structure methods remains largely undeveloped, however. There is a prospect, moreover, that further development of the present general-structure complementarity methods will lead to substantial improvements in their efficiency and range of applications. Due to the probable importance of complementarity methods in the development of new algorithms, the research program could pursue several major topics in this area.

[1] Linear Complementarity. Linear complementarity problems arise in linear and quadratic programming and in 2-person games, and they are a basic component of nonlinear programs and n-person games (see Cottle (1964), (1967), (1968a, b, c), (1970), (1971a, b, c), Eaves (1971a, b), and Lemke (1964), (1965)). In this area the research program could concentrate on the development and testing of methods which exploit the special structure of quadratic programs, especially ones of the large size and structure arising in major-system optimization problems [Beale (1967)].

[2] Nonlinear Complementarity. Nonlinear complementarity problems (see Cottle (1966), Eaves (1971d), Karamardian (1966), (1971) and Lemke (1970)) arise in general nonlinear programs and n-person games ($n \geq 3$). Normally such problems are most efficiently handled via linear or quadratic approximations. However, there is a variety of important nonlinear problems arising in practice whose special structure can be exploited to obtain more efficient procedures. The principal devices here are (a) the use of duality theory to obtain simpler dual problems or to pre-optimize subproblems of a larger system, and (b) the design of special complementarity algorithms which take advantage of the special structure. Both of these approaches could be pursued in the research program. A major class of practical problems which would be investigated are the pooling or the pre-processor problems. A major goal would be to convert systems of allocation and pooling problems into equivalent systems all of

one type or the other.

[3] Computation of Equilibria and Fixed-Points. One of the major outgrowths of complementarity methods has been the development for the first time of practical numerical methods for the computation of systems equilibria and fixed-points of mappings. (See Eaves (1970), (1971c, e, f), (1972), Freidenfelds (1971), Kuhn (1968), Scarf (1967a, b), (1969), (1972), Rosenmuller (1971), and Wilson (1971), (1972).) The advent of these methods opens the possibility of computing directly the equilibria of chemical, biological, and physical systems, and equilibria n-person games, rather than via the awkward approximation methods normally used. Moreover, it raises the possibility of a unified body of theory and computational methods (since, for example, convex programming problems can be shown to be equivalent to finding the fixed points of certain related mappings, and system equilibria are normally characterized either via the fixed-points of the equations of disequilibria or in terms of minimizing a measure of the loss from disequilibrium). The research program could pursue the further development of complementarity methods (including methods based on primitive sets and simplical subdivisions) for such problems with particular emphasis on the development of practical methods for computing the equilibria of large systems.

D. Combinatorial Problems and Integer Programming with Special Structure. The fundamental feature of many systems optimization problems is their combinatorial character. This may occur either because the problem has a special network structure or because it has discrete decision variables, so that a huge number of combinations must be considered. As Fulkerson (1966) discusses, such combinatorial problems arise in a wide variety of contexts. These problems sometimes can be solved, of course, but only by developing clever algorithms which exploit their special structure. Therefore, algorithmic development in this area will be one of the major research activities of a System Optimization Laboratory.

Probably the most important combinatorial problem for systems optimization is the integer programming problem [Gomory (1963)]. One reason is that so many optimization problems would be linear programs except that the decision variables make sense only with integer values (e.g., see Cushing (1969)), and so they become integer linear programs instead. In addition, it is possible to reformulate a number of important but difficult (indeed seemingly impossible) problems of a nonlinear, nonconvex and combinatorial character as mixed integer linear programming problems (see Dantzig (1963)). Another important reason is that many large-scale mathematical programming systems include subproblems which are integer programs, so that decomposition methods for such systems (e.g., see Benders (1962)) would need an integer programming algorithm as a subroutine.

Because of these considerations particular emphasis will be placed on algorithmic development for integer programming. This has been an area of substantial research for over a decade, and significant progress is being made (e.g., see Hillier (1969a), Balinski and Spielberg (1969), and Geoffrion and Marsten (1972)). Unfortunately, the problem is very difficult, and the efficiency of the available algorithms does not remotely approach that of the simplex method for the linear programming problem. Therefore, the main thrust of this research could be the development of special-purpose algorithms for important classes of integer programming problems in order to exploit special structure. Thorough testing and evaluation could be conducted, which would necessitate a major programming effort, so the resources and long-range continuity of the Systems Optimization Laboratory would play a vital role in carrying out this development beyond an initial stage. Decomposition methods for mixed integer programming systems could also be investigated. Another part of this research program would involve developing special-purpose heuristic procedures (see Hillier (1969b)) for obtaining good approximate solutions for large-scale integer programming systems having various common kinds of special structure.

E. <u>Further Development of a Macro-Language for Mathematical Programming.</u> Commerical codes for solving mathematical programming problems typically involve about 200,000 assembly-language instructions. One can anticipate that efficient commerical programs for solving structured systems optimization problems will be an order of magnitude more complex. In order for such programs to be developed and maintained, the language in which they are written must be highly readable and easy to modify. This is the purpose of the new MPL [Mathematical Programming Language] now under development. The continuation of this work could be one of the major projects of a Laboratory.

MPL is a high-level user-oriented programming language intended particularly for developing, testing, and communicating mathematical algorithms (see Dantzig, et al. (1970)). It is being developed to provide a language for mathematical algorithms that will be easier to write, to read, and to modify than currently available computer languages such as FORTRAN, ALGOL, PL/1, APL .

The need for a highly readable mathematically-based computer language has been apparent for some time. Generally speaking, standard mathematical notation in a suitably algorithm-like structure appears best for this purpose, since most researchers are familiar with the language of mathematics. Therefore, MPL closely parallels the vernacular of applied mathematics. An important area of application of MPL is for the development and testing of algorithms for systems optimization problems. To date, many methods have been proposed for solving such problems, but few have been experimentally tested and compared because of the high cost and the long time it takes to program them, and because it is difficult to debug and to modify them quickly after they are written. It is believed that highly readable programs would greatly facilitate experimentation with these proposed methods and would shorten the time until they can be used in practice. Thus, the development of a sophisticated version of MPL will provide a vital tool for the Systems Optimization Laboratory, as well as for other researchers.

As pointed out by William Orchard-Hays many other

special purpose languages beside MPL would be required as basic research tools. There is a need to have special language for Job Control, Computer Control, Matrix Generation, Procedure Programming (e.g., MPL or APL); languages for File Mechanisms; languages for organizing the entire system of computation.

. . . .

To summarize: Large-Scale Optimization requires laboratories where a large number of test models, computer programs, and special "tools" to aid in developing variants of existing techniques, are assembled in a systematic way. Only this way can one hope to model and solve the host of pressing total system problems that the world faces today.

REFERENCES

1. Abadie, J. M. and A. C. Williams (1963), "Dual and Parametric Methods in Decomposition", in Recent Advances in Math. Prog., R. Graves and P. Wolfe (eds.), McGraw-Hill, New York, pp. 149-158.

2. Aronofsky, Julius S. (ed.) (1969), Progress in Operations Research (Relationship Between Operations Research and the Computer), Vol. III, Wiley, New York.

3. Bartels, R. H., (1971), "A Stabilization of the Simplex Method", Numerische Mathematik, Vol. 16, pp. 414-434.

4. Bartels, R. H. and G. H. Golub (1969), "The simplex method of linear programming using LU decomposition", Comm. ACM, Vol. 12, pp. 266-268.

5. Balas, Egon, (1966), "An Infeasibility - Pricing Decomposition Method for Linear Programs", Operations Research, Vol. 14, pp. 843-873.

6. Balinski, M. L. and K. Spielberg, (1969), "Methods for Integer Programming: Algebraic, Combinatorial, and Enumerative", Ch. 7, pp. 195-292, in Progress in Operations Research (Relationship Between Operations Research and the Computer), J. S. Aronofsky (ed.), Wiley, New York.

7. Balinski, M. L. (1966), "On Some Decomposition Approaches in Linear Programming", and "Integer Programming", The University of Michigan Engineering Summer Conferences.

8. Baumol, W. J. and T. Fabian, (1964), "Decomposition, Pricing for Decentralization and External Economies", Management Science, Vol. 11, No. 1, pp. 1-32.

9. Beale, E. M. L., (ed.)(1970), Applications of Mathematical Programming Techniques, American Elsevier, New York.

10. Beale, E. M. L. (1967), "Decomposition and Partitioning Methods for Nonlinear Programming", in Non-Linear Programming, J. Abadie (ed.), North-Holland, Amsterdam.

11. Beale, E. M. L., (1963), "The Simplex Method Using Pseudo-Basic Variables for Structured Linear Programming Problems", from Recent Advances in Math. Prog., R. Graves and P. Wolfe (eds.) McGraw-Hill, New York, pp. 133-148.

12. Benders, J. F., (1962), Partitioning Procedures for Solving Mixed-Variables Programming Problems", Numerische Mathematik, Vol. 4, pp. 238-252.

13. Bennett, J. M. (1966), "An Approach to Some Structured Linear Programming Problems", Operations Research, 14, 4 (July-August), pp. 636-645.

14. Broise, P., P. Huard, and J. Sentenac, (1968), Décomposition des Programmes Mathématiques, Monographies de Recherche Opérationnelle, Dunod, Paris.

15. Cottle, R. W. and J. Krarup (eds.), (1971), Optimization Methods for Resource-Allocation, Proceedings of the NATO Conference, Elsinore, Denmark. English University Press, London, 1973.

16. Cottle, R. W. (1964), "Note on a Fundamental Theorem in Quadratic Programming", SIAM J. Appl. Math., 12, pp. 663-665.

17. Cottle, R. W. (1966), "Nonlinear Programs with Positively Bounded Jacobians", SIAM J. Appl. Math., 14, 1, pp. 147-158.

18. Cottle, R. W. and G. B. Dantzig (1967), "Positive (Semi-) Definite Programming", in Nonlinear Programming, North-Holland, J. Abadie (ed.), pp. 55-73, Amsterdam.

19. Cottle, R. W. and G. B. Dantzig (1968a), "Complementary Pivot Theory of Mathematical Programming", Lin. Alg. and Its Applns., 1, pp. 103-125.

20. Cottle, R. W. (1968b), "The Principal Pivoting Method of Quadratic Programming", in Mathematics of the Decision Sciences, Part I, American Mathematical Society, G. B. Dantzig and A. F. Veinott, Jr. (eds.), Providence, R. I., pp. 144-162.

21. Cottle, R. W. (1968c), "On a Problem in Linear Inequalities", J. London Math. Soc., 42, pp. 378-384.

22. Cottle, R. W. and W. C. Mylander (1970), "Ritter's Cutting Plane Method for Non-convex Quadratic Programming", in Integer and Nonlinear Programming, North-Holland, J. Abadie (ed.), Amsterdam, pp. 257-283.

23. Cottle, R. W. and A. F. Veinott, Jr. (1971a), "Polyhedral Sets Having a Least Element", Math Prog. Vol. 3, pp. 238-249.

24. Cottle, R. W. (1971b), "Solution to G. Maier's Problem on Parametric Linear Complementarity Problems", submitted to SIAM Review.

25. Cottle, R. W. (1971c), "Monotone Solutions of the Parametric Linear Complementarity Problem", Technical Report No. 71-19, Department of Operations Research, Stanford University, submitted to Math. Prog.

26. Cushing, B. E., (1969), "The Application Potential of Integer Programming", The Journal of Business, Vol. 42, pp. 457-467.

27. Dantzig, G. B., (1963), Linear Programming and Extensions, Princeton University Press, Princeton, N. J.

28. Dantzig, G. B., (1960), "On the Significance of Solving Linear Programming Problems with Some Integer Variables", Econometrica, Vol. 28, pp. 30-44.

29. Dantzig, G. B. and P. Wolfe, (1960), "Decomposition Principle for Linear Programming Problems", Operations Research, Vol. 8, pp. 101-110.

30. Dantzig, G. B. and R. M. Van Slyke, (1964), "Generalized Upper Bounded Techniques for Linear Programming - I", Proceedings IBM Scientific Computing Symposium, Combinatorial Problems, March 16-18, pp. 249-261.

31. Dantzig, G. B. and R. M. Van Slyke, (1967), "Generalized Upper Bounded Techniques for Linear Programming - II", J. Computer and System Sciences, Vol. 1, pp. 213-226.

32. Dantzig, G. B., S. Eisenstat, T. Magnanti, S. Maier, M. McGrath, V. Nicholson, C. Riedl, (1970), "MPL-Mathematical Programming Language-Specification Manual", Stanford University Computer Science Department Technical Report, STAN-DS-70-187.

33. Dantzig, G. B., (1963), "Compact basis triangularization for the simplex method" in R. Graves and P. Wolfe (eds.), Recent Advances in Math. Prog., McGraw-Hill, New York, pp. 125-132.

34. Dantzig, G. B., (1955), "Upper Bounds, Secondary Constraints and Block Triangularity in Linear Programming", Econometrica, Vol. 23, No. 2, pp. 174-183.

35. Dantzig, G. B., (1955), "Optimal Solution of a Dynamic Leontief Model with Substitution", Econometrica, Vol. 23, No. 3, pp. 295-307.

36. Dantzig, G. B., (1955), "Linear Programming Under Uncertainty", Management Science, Vol. 1, pp. 197-206.

37. Dantzig, G. B., (1966), "Linear Control Processes and Mathematical Programming", SIAM J. Control, Vol. 4, No. 1, pp. 56-60.

38. Dantzig, G. B., D. R. Fulkerson, and S. Johnson, (1954), "Solution of a Large Scale Traveling Salesman Problem", Operations Research, Vol. 2, No. 4, pp. 393-410.

39. Dantzig, G. B., R. Harvey, R. McKnight, (1964), "Updating the Product Form of the Inverse for the Revised Simplex Method", Operations Research Center, University of California, Berkeley.

40. Dantzig, G. B. and R. M. Van Slyke (1967), "Generalized Upper Bounding Techniques", J. Computer System Science, Vol. 1, pp. 213-226.

41. Dzielinski, P. and R. E. Gomory (1963) (abstract), "Lot size Programming and the Decomposition Principle", Econometrica, Vol. 31, p. 595.

42. Eaves, B. C. (1970), "An Odd Theorem", Proc. of Amer. Math. Soc., 26, 3, pp. 509-513.

43. Eaves, B. C. (1971a), "The Linear Complementarity Problem", Management Science, 17, 9, pp. 612-634.

44. Eaves, B. C., (1971b), "On Quadratic Programming", Management Science, 17, 11, pp. 698-711.

45. Eaves, B. C., (1971c), "Computing Kakutani Fixed Points", SIAM J. Appl. Math., 21, 2, pp. 236-244.

46. Eaves, B. C., (1971d), "On the Basic Theorem of Complementarity", Math. Prog., 1, 1, pp. 68-75.

47. Eaves, B. C., (1971e), "Homotopies for Computation of Fixed Points", Math. Prog. Vol. 3, pp. 1-22.

48. Eaves, B. C., and R. Saigal (1971f), "Homotopies for Computation of Fixed Points on Unbounded Regions", Math. Prog. Vol. 3, pp. 225-237.

49. Eaves, B. C., (1972), "Polymatrix Games with Joint Constraints", Department of Operations Research, Stanford University, submitted to SIAM J. Appl. Math.

50. Electricité de France, Direction des études et recherches, "Programmation Linéaire, Méthode de Dantzig et Wolfe, Programme Expérimental", Paris.

51. Elmaghraby, S. E., (1970), The Theory of Networks and Management Science, Parts I, II," Management Science, Vol. 17, pp. 1-34, B-54-71.

52. Ford, L., Jr., and D. R. Fulkerson, (1958), "Suggested Computation for Maximal Multi-Commodity Network Flows", Management Science, Vol. 5, No. 1, pp. 97-101.

53. Forrest, J. J. H., and J. A. Tomlin, "Updating triangular factors of the basis to maintain sparsity in the product form simplex method", in Cottle and Krarup (forthcoming).

54. Freidenfelds, J., (1971), "A Fixed-Point Algorithm and Almost-Complementary Sets", Technical Report No. 71-3, Department of Operations Research, Stanford University. In addition, "Fixed-Point Algorithms and Almost-Complementary Sets", Ph. D. Dissertation, Department of Operations Research, Stanford University.

55. Frisch, R., (1962), "Tentative Formulation of the Multiplex Method for the Case of a Large Number of Basic Variables", Institute of Economics, University of Oslo.

56. Fulkerson, D. R., (1966), "Flow Networks and Combinatorial Operations Research", American Mathematical Monthly, Vol. 73, pp. 115-138.

57. Gale, D., (1966), "On Optimal Development in a Multi-sector Economy", Operations Research Center, 1966-11, University of California, Berkeley.

58. Gass, S. I., (1966), "The Dualplex Method for Large-Scale Linear Programs", Operations Research Center, 1966-15, University of California, Berkeley, Ph. D. Thesis.

59. Geoffrion, A. M., (1970), "Elements of Large-Scale Mathematical Programming", Management Science, Vol. 16, pp. 652-691.

60. Geoffrion, A. M., (1971), "Optimal Distribution System Design", Proceedings of the NATO Conference, Optimization Methods for Resource Allocation, R. W. Cottle and J. Krarup, (eds), Elsinore, Denmark, English University Press, London, 1973.

61. Geoffrion, A. M. and R. E. Marsten, (1972), "Integer Programming Algorithms: A Framework and State-of-the-Art Survey", Management Science, Vol. 18, (forthcoming).

62. Gill, P. E. and W. Murray, (1971), "A Numerically Stable Form of the Simplex Algorithm", Technical report No. Maths 87, National Physical Laboratory, Teddington.

63. Gill, P. E., G. H. Golub, W. Murray and M. A. Saunders, "Methods for Modifying Matrix Factorizations", to be published.

64. Gilmore, P. C., and R. E. Gomory, (1965), "Multistage Cutting Stock Problems of Two and More Dimensions", Operations Research, Vol. 13, pp. 94-120.

65. Gilmore, P. C., and R. E. Gomory, (1961), "A Linear Programming Approach to the Cutting Stock Problem", Operations Research, Vol. 9, pp. 849-859.

66. Glassey, C. R., (1971), "Dynamic Linear Programs for Production Scheduling", Operations Research, Vol. 19, pp. 45-56.

67. Gomory, R. E., (1963), "Large and Non-Convex Problems in L. P.", Proc. Sympos. Appl. Math. 15, pp. 125-139.

68. Gomory, R. E., and T. C. Hu, (1962), "An Application of Generalized Linear Programming to Network Flows", SIAM Journal, Vol. 10, No. 2, pp. 260-283.

69. Graves, R. L., P. Wolfe, (eds.) (1963), Recent Advances in Mathematical Programming, McGraw-Hill, New York

70. Haley, R. B., (1966), "A General Method of Solution for Special Structure Linear Programmes", Operational Research Quarterly, Vol. 17, No. 1, pp. 83-90.

71. Heesterman, A. R. G., and J. Sandea, (1965), "Special Simplex Algorithm for Linked Problems", Management Science, 11, 3, pp. 420-428.

72. Hellerman, E., (1965), "The Dantzig-Wolfe Decomposition Algorithm as Implemented on a Large-Scale (Systems Engineering) Computer", presented at Modern Techniques in the Analysis of Large-Scale Engineering Systems.

73. Hu, T. C., (1963), "Multi-Commodity Network Flows", Operations Research, Vol. 11, pp. 344-360.

74. Hillier, F. S., (1969a), "A Bound-and-Scan Algorithm for Pure Integer Linear Programming with General Variables", Operations Research, Vol. 17, pp. 638-679.

75. Hillier, F. S., (1969b), "Efficient Heuristic Procedure for Integer Linear Programming with an Interior", Operations Research, Vol. 17, pp. 600-637.

76. Kutcher, G. P., "On Decomposing Price-Endogeneous Models", in L. M. Gorens and A. S Manne (eds.), Multi-level Planning: Case Studies in Mexico, North Holland, Amsterdam, forthcoming, Ch. V. 2.

77. Karamardian, S., (1966), "Duality in Mathematical Programming", Operations Research Center, University of California, Berkeley, or Parts 1 and 2 of "The Nonlinear Complementarity Problem with Applications", JOTA, 4, 2, and 3 (1969), pp. 87-98, pp. 167-181.

78. Karamardian, S., (1971), "The Complementarity Problem", Graduate School of Administration and Department of Mathematics, University of California, Irvine, to appear in Math. Prog.

79. Kuhn, H., (1968), "Simplicial Approximation of Fixed Points", Proc. Nat. Acad. Sci. U. S. A., 61, pp. 1238-1242.

80. Kron, G., (1963), Diakoptics, The Piecewise Solution of Large-Scale Systems, Macdonald and Co., London.

81. Lasdon, L. S., (1970), "Optimization Theory for Large Systems", The Macmillan Company, New York.

82. Lasdon, L. S., (1971, "Uses of Generalized Upper Bounding Methods in Production Scheduling", in R. W. Cottle and J. Krarup (eds), Optimization Methods for Resource-Allocation, Proceedings of the NATO Conference, Elsinore, Denmark, English University Press, London, 1973.

83. Lemke, C. E., (1965), "Bimatrix Equilibrium Points and Mathematical Programming", Management Science 11, 7, pp. 681-689.

84. Lemke, C. E., (1970), "Recent Results in Complementarity Problems", in Nonlinear Programming, J. B. Rosen, O. Mangasarian and K. Ritter, (eds.), Academic Press, New York, pp. 349-384.

85. Lemke, C. E. and J. T. Howson, Jr., (1964), "Equilibrium Points of Bimatrix Games", SIAM J. Appl. Math., 12, pp. 413-423.

86. Reid, J. K., (ed.) (1971), "Large Sparse Sets of Linear Equations", Academic Press, New York.

87. Ritter, K., (1967), "A Decomposition Method for Linear Programming Problems with Coupling Constraints and Variables", Math. Research Center, The University of Wisconsin #739. Published with other works as: M. D. Grigoriadis and K. Ritter, "A decomposition method for structured linear and non-linear programs", J. Comput. Systems Sci., 3 (1969), pp. 335-360.

88. Rosen, J. B., (1964), "Primal Partition Programming for Block Diagonal Matrices", Numerische Math., 6, 3, pp. 250-264.

89. Rosenmüller, J., (1971), "On a Generalization of the Lemke-Howson Algorithm to Noncooperative N-Person Games", SIAM J. Appl. Math., 21, 1, pp. 73-79.

90. Saunders, M. A., (1972), "Large-scale linear programming using the Cholesky factorization", Technical Report No. CS-252, Computer Science Department, Stanford University.

91. Scarf, H., (1967a), "The Core of an N-Person Game", Econometrica, 35, pp. 50-69.

92. Scarf, H., (1967b), "The approximation of Fixed Points of a Continuous Mapping", SIAM J. Appl. Math., 15, pp. 1328-1343.

93. Scarf, H., and T. Hansen, (1969), "On the Applications of a Recent Combinatorial Algorithm", Discussion Paper No. 272, Cowles Foundation, Yale University.

94. Scarf, H., in collaboration with T. Hansen (1972), The Computation of Economic Equilibria, Cowles Foundation, Yale University, to be published.

95. Tomlin, J. A., (1971), "Modifying triangular factors of the basis in the simplex method", presented at the Symposium on Sparse Matrices and their Applications, IBM Thomas J. Watson Research Center, Yorktown Heights, New York.

96. Van Slyke, R. M. and R. Wets, (1969), "L-Shaped Linear Programs with Applications to Optimal Control and Stochastic Programming", SIAM J. Appl. Math., 17, pp. 638-663.

97. Van Slyke, R. and R. Wets, (1966), "Programming Under Uncertainty and Stochastic Optimal Control", SIAM J. Control, Vol. 14, No. 1, pp. 179-193.

98. Varaiya, P., (1966), "Decomposition of Large-Scale Systems", SIAM J. Control, Vol. 4, No. 1, pp. 173-179.

99. Wagner, H. M., (1957), "A Linear Programming Solution to Dynamic Leontief Type Models", Management Science, Vol. 3, No. 3, pp. 234-254.

100. Wilde, D. J., (1963), "Production Planning of Large Systems", Chemical Engineering Progress, Vol. 59, pp. 46-51.

101. Williams, J. D., (1960), "The Method of Continuous Coefficients, Parts I and II", Report No. ECC 60.3, Socony.

102. Williams, A. C., (1962), "A Treatment of Transportation Problems by Decomposition", SIAM Journal, Vol. 10, No. 1, pp. 35-48.

103. Willoughby, R. A., (ed.) (1969), Proceedings of the Symposium on Sparse Matrices and Their Applications, IBM Thomas J. Watson Research Center, Yorktown Heights, New York.

104. Wilson, R. B., (1971), "Computing Equilibria of N-Person Games", SIAM J. Appl. Math., 21, 1, pp. 80-87.

105. Wilson, R. B., (1972), "Computation of Equilibria of Two-Person Games in Extensive Form", Graduate School of Business, Stanford University, to appear in Management Science.

106. Wismer, D. A., (ed.) (1971) "Optimization Methods for Large-Scale Systems... With Applications", McGraw-Hill, New York.

107. Wolfe, P., (1961) "Accelerating the Cutting Plane Method for Non-linear Programming", SIAM Journal, Vol. 9, No. 3, pp. 481-488.

108. Wolfe, P., and G. B. Dantzig, (1962), "Linear Programming in a Markov Chain", Operations Research, 10, pp. 702-710.

109. Wood, M. K., and G. B. Dantzig (1947), "The Programming of Interdependent Activities", in T. C. Koopmans, Activity Analysis of Production and Allocation, Wiley, New York, 1951; Also Econometrica Vol. 17, Nos. 3 and 4, July-October, 1949, pp. 193-199.

110. Zschau, E. V. W., (1967), "A Primal Decomposition Algorithm for Linear Programming", Graduate School of Business, Stanford University.

FOOTNOTES

1. Parts of the material in this paper were drawn from a draft of a proposal to establish such a laboratory at Stanford University, prepared by R. Cottle, B. C. Eaves, G. H. Golub, F. S. Hillier, A. S. Manne, D. J. Wilde, R. B. Wilson and the author.

2. This paper will also appear in Optimization Methods for Resource Allocation, English University Press, London, 1973.

Department of Operations Research
Stanford University
Stanford, California. 94305

A Markov Decision Problem

ERIC V. DENARDO

A much-studied <u>Markov decision model</u> is first described. A system is observed at equally-spaced epochs numbered $0, 1, 2, \ldots$. At each epoch the system is observed to occupy one of N states, which are numbered 1 through N . Each state i has associated with it a finite non-empty decision set D_i. Whenever state i is observed, some decision k in D_i must be selected. Suppose state i is observed at epoch n and decision k is selected. Reward R_i^k is earned immediately, and the probability that state j is observed at epoch $n+1$ is P_{ij}^k . This model is stationary in that nothing depends on n. Transitions are presumed to occur with probability 1, so that

$$\sum_{j=1}^{N} P_{ij}^{k} = 1 \qquad \text{all } i, \text{ all } k .$$

A (stationary) <u>policy</u> δ is a decision procedure that specifies for each state i a decision $\delta(i)$ from D_i . The <u>policy space</u> Δ is the set of all such policies. This means that selecting a particular decision for state i in no way restricts our choice of decision for any other states. Suppose state i is observed and policy δ is in effect. Reward $R_i^{\delta(i)}$ is earned immediately, and the probability that state j will next be observed is $P_{ij}^{\delta(i)}$. These numbers assume their respective positions in the $N \times 1$ reward vector R^δ and the

$N \times N$ transition matrix P^δ. For instance, the ith element of R^δ is the reward for observing state i and selecting decision $\delta(i)$.

This model is analyzed here under several performance criteria, each of which involves the limiting behavior of the income earned at the first n epochs, as n approaches infinity. Throughout, we assume that each transition matrix P^δ has one ergodic chain, which must be acyclic. Transient states are allowed, and they can vary with the policy.

This assumption is invoked for several reasons. In a sense, it is neither too easy nor too hard. As to why periodicities are excluded, Feller [6, page 356] offers the apt comment, "The modifications required for periodic chains are rather trite, but the formulations become unpleasantly involved." Multiple chains are excluded because they severely complicate parts of the development and sometimes require more involved solution techniques. Even so, the line of attack used here goes a long way toward analyzing the general multi-chain case, with periodicities allowed. The mathematics of this chapter would have been much easier if we had eliminated the possibility of transient states. However, analysis of the model without transient states provides precious little insight into how to analyze the more general situations.

Our assumption has the effect of excluding periodicities and presuming that at least one particular state is recurrent, no matter what the policy. A reasonably wide class of decision problems exhibits these features. Though we shall not go into details, several of our methods also work when the optimal policies have these features and, possibly, the suboptimal ones do not. On the other hand, the assumption of no transient states would be satisfied by virtually no examples.

The text has three main sections. In the first two of these, attention is restricted to stationary policies. The income generating characteristics of a particular policy are first provided, and then the question of computing an optimal stationary policy is considered. The third main section introduces non-stationary policies and determines certain operating characteristics of the optimal (non-stationary) policy for a very long planning horizon.

The text itself is nearly self-contained and serves as an introduction to Markov decision problems, as well as developing some new results. References are included in the text only for the facts not developed there. A final section contains citations.

A Markov Process with Rewards

As a prelude to the optimization problem, we first examine the income generation characteristics of a fixed policy. While doing so, we can and shall simplify the notation by deleting the k in P_{ij}^k and in R_i^k. So P is the $N \times N$ matrix† whose ijth element is P_{ij}, and R is the $N \times 1$ vector with ith element R_i.

A square matrix, like P, with nonnegative entries and row sums equal to 1 is called a <u>stochastic</u> matrix. Suppose state i is observed at epoch 0. Then P_{ij} is the probability that state j is observed at epoch 1. Also, the probability that state j is observed at epoch 2 is $\Sigma_k P_{ik} P_{kj}$, which is the ijth entry of the matrix P^2. Similarly, the ijth element of P^n is the probability that state j is observed at epoch n.

Again, suppose state i is observed at epoch 0. Reward R_i is earned immediately. The expectation of the income earned at epoch 1 is $\Sigma_{j=1}^N P_{ij} R_j$, which is the ith component of the vector PR. Similarly, the expectation of the reward earned at epoch n is the ith component of the

†As a memory aid, certain conventions are followed throughout. Matrices and vectors are usually capitalized, and Q_{ij} and R_i are elements of the matrix Q and vector R. Also, I is the $N \times N$ identity matrix, 0 is often a matrix of 0's of appropriate dimension, and 1 is the $N \times 1$ vector of 1's. When Q is square, $Q^2 = QQ$, etc. Finally, when the range of a variable is left unspecified, its entire range is intended; e.g., $\Sigma_j P_{ij} = \Sigma_{j=1}^N P_{ij}$ and $\max_i R_i = \max_{1 \le i \le N} R_i$.

vector P^nR. Consider the $N \times 1$ vector V^n, defined by

$$V^n = R + PR + \dots + P^{n-1} R$$

we have just observed that the ith component of the vector V^n is the expectation of the income earned at epochs 0 through n - 1, providing state i is observed at epoch 0. This equation is ample testament to the utility of matrix notation. Since R postmultiplies each term in the above equation, it can be factored.

(1) $$V^n = \left(\sum_{i=0}^{n-1} P^i \right) R$$

The reward vector R plays a secondary role; interest centers upon the sum of the first n powers of the stochastic matrix P, especially as the planning horizon n becomes large. Under our assumption, there exist $N \times N$ matrices P^* and Y such that

(2) $$\sum_{i=1}^{n-1} P^i = nP^* + Y + o(1) \qquad \text{as } n \to \infty ,$$

where, as is usual in mathematics, o(1) is short for a matrix of error terms, each of which approaches zero as the limit is approached. So, (2) partitions the sum of the first n powers of P into a term that grows linearly in n, a constant term, and an error term that approaches zero as $n \to \infty$. Equations (1) and (2) lead us to define $N \times 1$ vectors g and w by

(3) $$g = P^*R \qquad w = YR .$$

Substitute (2) and (3) into (1) and note that $o(1)R = o(1)$. The result is

(4) $$V^n = ng + w + o(1) \qquad \text{as } n \to \infty .$$

Thus, as n becomes large, the ith component of V^n approaches the straight-line $ng_i + w_i$ having slope g_i and intercept w_i with the ordinate. This is shown diagrammatically in Figure la, which plots V_1^n versus n. The number g_i is called the gain-rate for state i, and w_i is called its bias.

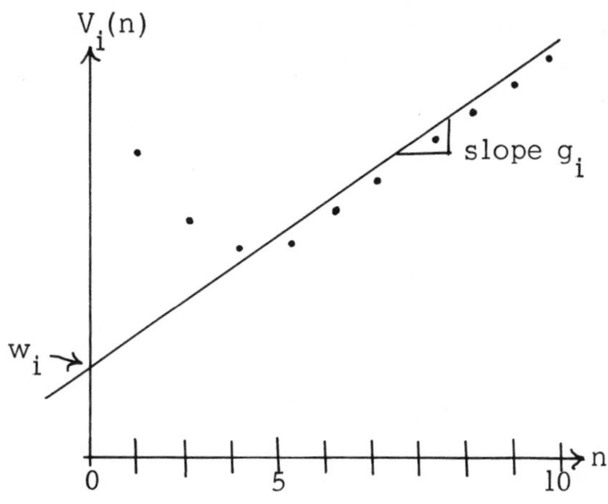

Figure la: Asymptotic behavior of $V_i(n)$.

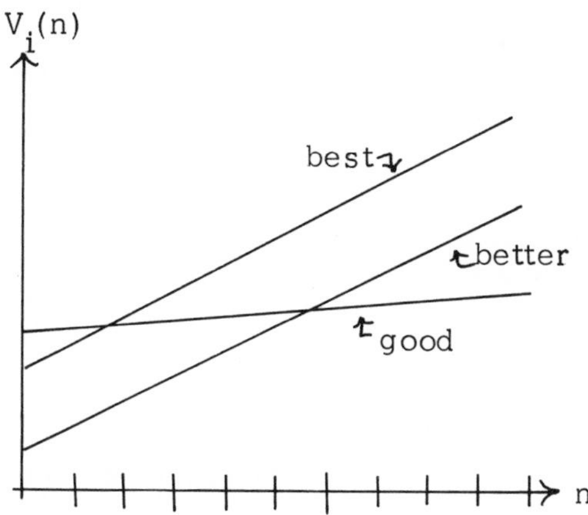

Figure lb: Asymptotic lines for 3 policies.

Suppose the planning horizon is long and the decision-maker is offered a choice of policies. Since the income-generation behavior of each approaches a straight line, he can compare lines, rather than policies. Presumably, he would prefer a higher slope and, among lines of equal slope, a higher bias. That preference system yields him maximum income for all sufficiently long planning horizons, barring ties. This situation is depicted in Figure 1b.

Theorems 1 and 2, which follow, will be used to verify (2).

THEOREM 1.

(a) There exists exactly one $1 \times N$ vector s such that $s = sP$ and $\sum_{i=1}^{N} s_i = 1$.

(b) $P^n \to P^*$, where P^* is a stochastic matrix each of whose rows is s.

(c) $P^* = PP^* = P^*P = P^*P^*$.

(d) Let $S = P - P^*$. For each integer $i \geq 1$,

$$S^i = P^i - P^* . \tag{5}$$

The standard proof of Theorem 1 is contained, among other places, in Kemeny and Snell [8]. A proof based on contraction mappings will be found in Denardo [5].

In order to use Theorem 1, we add and subtract nP^* from the sum of the first n powers of P. This results in the identity

$$\sum_{i=0}^{n-1} P^i = nP^* + \sum_{i=0}^{n-1} (P^i - P^*) . \tag{6}$$

Substitute (5) into (6), and account for the case $i = 0$.

$$\sum_{i=0}^{n-1} P^i = nP^* + \sum_{i=0}^{n-1} S^i - P^* \tag{7}$$

Equation (5) and Part (b) of Theorem 1 assure $S^n \to 0$ as $n \to \infty$. Hence, the sum on the right-hand side of (7) is the matrix analogue of the partial sum of a convergent geometric series. Theorem 2 gives matrix analogues of the formulas for the partial and complete sums of geometric series, where $(I - S)^{-1}$ is the analogue of the reciprocal.

THEOREM 2.

$$\sum_{i=0}^{\infty} S^i = (I - S)^{-1} \tag{8}$$

$$\sum_{i=0}^{n-1} S^i = (I - S^n)(I - S)^{-1} . \tag{9}$$

A proof of (8) can be found in Kemeny and Snell [8]. Equation (9) can be established similarly. Theorem 2 depends only on the fact that $S^n \to 0$, though we shall only apply it for $S = P - P^*$. Equation (8) facilitates definition of the <u>fundamental</u> <u>matrix</u> Z by

$$Z = \sum_{i=0}^{\infty} (P - P^*)^i = (I - P + P^*)^{-1} . \tag{10}$$

Substitute (9) into (7), yielding

$$\begin{aligned} \sum_{i=0}^{n-1} P^i &= nP^* + (I - S^n)Z - P^* \\ &= nP^* + Z - P^* - (P^n - P^*)Z , \end{aligned} \tag{11}$$

the last from (5). Since $P^n \to P^*$, equation (11) verifies (2) when we set

$$Y = Z - P^* . \tag{12}$$

Not only does (11) verify (2), it provides an explicit formula for the error term.

Certain facts about Y are now demonstrated. Equations (10) and (5) imply

$$Y = \sum_{i=0}^{\infty} (P^i - P^*) . \tag{13}$$

Premultiply and postmultiply (13) by P^* to see that

(14) $$P^*Y = YP^* = 0 .$$

Equation (14) allows us to rearrange (11) as

(15) $$\sum_{i=0}^{n-1} P^i = nP^* + Y - P^nY .$$

Since $P^nY \to P^*Y = 0$, equation (15) also realizes the expansion sought in (2). Postmultiply (13) by 1 to see that

(16) $$Y1 = 0 .$$

Equation (14) and part (c) of Theorem 1 imply

$$Y(I - P) = Y(I - P + P^*)$$

$$= (Z - P^*)(I - P + P^*)$$

(17) $$= I - P^* .$$

Computation:

Attention now turns to the computation of P^* and Y. Each row of P^* equals s. Part (a) of Theorem 1 states that s is the unique solution of the equation system

(18) $$s(I - P) = 0 , \qquad s1 = 1 .$$

Equation (18) is actually a system of $N + 1$ equations in N unknowns. Since it specifies s uniquely, it must contain exactly one redundancy. Since P is stochastic, the columns of $(I - P)$ sum to zero; i.e.,

(19) $$(I - P)1 = 0 .$$

So, any one of the columns of $I - P$ can be eliminated. Let B be the $N \times N$ matrix obtained by replacing the first column of $I - P$ by 1. Then (18) and (19) combine to assure that s is the unique solution of

(20) $$sB = [1, 0, \ldots, 0] .$$

This means that B is invertible and that

(21) $$s = [1, 0, \ldots, 0] B^{-1} .$$

That is, s is the top row of B^{-1}.

So s is found by inverting B. Equation (10) indicates that Y can then be found by inverting $(I - P + P^*)$. We now show that the second inversion is superfluous. Consider the columns of the matrix YB. Equation (16) states that the first column of YB equals 0. Equation (17) assures that the 2nd through Nth columns of YB equal the corresponding columns of $I - P^*$. In summary, we have demonstrated

(22) $$YB = C ,$$

where C is the $N \times N$ matrix obtained from $(I - P^*)$ by replacing its first column with zeros. Postmultiply (22) by B^{-1}.

(23) $$Y = CB^{-1} .$$

Equation (23) obtains Y without a second inversion. This development is summarized as Theorem 3.

THEOREM 3. Let B be the matrix obtained by replacing the first column of $(I - P)$ with 1. Let C be the matrix obtained by replacing the first column of $(I - P^*)$ with 0. Then B is invertible, the top row of B^{-1} equals s, and $Y = CB^{-1}$.

<u>Remark:</u> The asymmetry in B and C are due to having replaced particular columns by 1 and 0. For a symmetric variant of Theorem 3, consider the matrix D whose ijth element is $\delta_{ij} - P_{ij} + 1/N$. An argument similar to the above

one verifies that D is invertible, that $S = 1^t D^{-1}$ and $Y = (I - P^*)D^{-1}$. We developed the asymmetric version because it is the one that occurs in linear programming and in policy iteration.

The policy evaluation equation:

Having computed P^* and Y, we can determine the gain rate g and bias w by postmultiplying by R, as (3) indicates. Even so, we shall develop a separate equation, from whose solution g and w can also be computed. This equation plays a prominent role in policy iteration and in linear programming.

Since all rows of P^* equal s, the gain rate does not vary from state to state. So we can define a scalar gain rate g by the equation $g = sR$ and replace g by $g1$. With $w = YR$, substituting (15) into (1) yields

$$V^n = ng1 + w - P^n w \,. \tag{24}$$

One can verify by induction on n or by direct substitution into (1) that

$$V^{n+1} = R + PV^n \,. \tag{25}$$

Equation (25) is one of the standard recursive relationships of dynamic programming. Substitute (24) on both sides of (25).

$$(n+1)g1 + w - P^{n+1}w = R + P[ng1 + w - P^n w] \,. \tag{26}$$

Recall that $P1 = 1$ and cancel four terms in (26).

$$g1 + w = R + Pw \,. \tag{27}$$

Equation (27) is called the policy evaluation equation. It can be obtained more quickly by observing that $YP = PY$ and then postmultiplying (17) by R. However, the method we detailed seems more intuitive. Equation (27) is actually a system of N equation in N + 1 unknowns, for which reason it cannot determine g and w uniquely. For any scalar m, replace w by $w + m1$ in (27); equality is preserved. Theorem 4 verifies that this is the only source of ambiguity.

THEOREM 4. Consider any scalar x and vector z such that

$$x1 + (I - P)z = R \quad . \tag{28}$$

Then $g = x$ and $w = z - 1(sz)$.

Remark: (28) is the policy evaluation equation, written in slightly different notation, with Pz transposed to the left-hand side.

PROOF: To obtain g and w on the right-hand side of (28), we shall premultiply it by s and Y , respectively. Since $s(I - P) = 0$, premultiplying (28) by s reduces it to

$$x + 0 = sR = g \quad .$$

Since $Y1 = 0$ and $Y(I - P) = I - P^*$, premultiplying (28) by Y reduces it to

$$0 + (I - P^*)z = YR = w \; .$$

Since $P^* = 1s$, the above implies $z - 1sz = w$, which completes the proof of Theorem 4 . ■

Two solution of (28) are of particular interest. The solution having $z_1 = 0$ reduces to

$$B \begin{pmatrix} g \\ z_2 \\ \vdots \\ z_N \end{pmatrix} = R \, , \tag{29}$$

which can readily be recovered from B^{-1}. Theorem 4 asserts that if $sz = 0$, then $z = w$, which verifies the following:

COROLLARY 1. g and w are the unique solution of (30).

$$gl + (I - P)w = R, \qquad sw = 0. \tag{30}$$

Optimization

We are now equipped to deal with the optimization problem. The decision k and policy δ re-emerge as superscripts on the data P_{ij}^k, R_i^k, P^δ and R^δ, as well as on the derived notation $g^\delta, w^\delta, s^\delta, Y^\delta, B^\delta, C^\delta$, etc.

Two Criteria:

It was suggested by Figure 1 that a decision-maker would prefer a higher gain-rate and, if two policies have the same gain-rate, a higher bias. These notions are inherent in the following definitions of g^* and w^*. Recall that Δ is the set of all policies.

$$g^* = \max \{g^\delta \mid \delta \in \Delta\}$$

$$w_i^* = \max \{w_i^\delta \mid \delta \in \Delta,\ g^\delta = g^*\} \quad i=1,\dots,N.$$

Policy π is called gain-optimal if $g^\pi = g^*$ and bias-optimal if, in addition, $w^\pi = w^*$. To show existence of a gain-optimal policy, simply note that g^δ is a scalar and that δ ranges over a finite set, namely Δ. However, a bias-optimal policy must attain w_i^* simultaneously for each i, so that its existence requires demonstration.

Policy Iteration:

Policy iteration is now used to compute a gain-optimal policy. As usual, the policy iteration routine in Figure 2 consists of alternating policy evaluation and policy improvement steps.

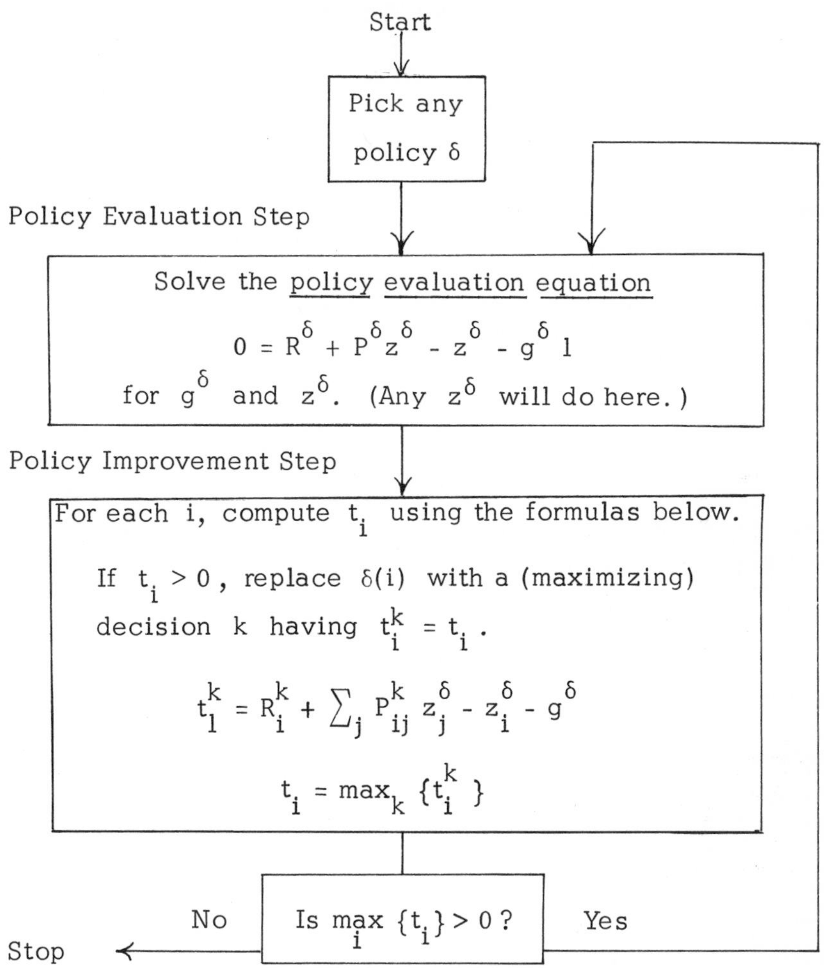

Figure 2: A policy iteration routine.

Remark: The policy evaluation equation can be solved by inverting B^{δ} and using (29). But, on successive iterations, the matrices requiring inversion may differ by only one or a few rows. When this occurs, it will be more efficient to update the old inverse than to reinvert the entire matrix, just as it is in linear programming. In fact, the inverse can even be stored in product form, as it itself is not ever required by the algorithm.

Remark: Compare the policy evaluation and policy improvement steps to see that $t_i^k = 0$ whenever $k = \delta(i)$. Consequently, t_i is never negative. The policy iteration routine only changes $\delta(i)$ when $t_i > 0$. Actually, one can pick any maximizing k , even when $t_i = 0$, but doing this invalidates Lemma 1, complicates the proof of convergence, and confounds the analogy with linear programming.

The following lemma has several uses, one of which concerns policy iteration. The lemma assures that each nonterminal repetition of the policy improvement step either increases the gain-rate or maintains it and increases the bias.

LEMMA 1: Consider any policy δ and any solution (g^δ, z^δ) of the policy evaluation equation. Suppose policy π satisfies (31) and (32).†

$$0 < t = R^\pi + (P^\pi - I)z^\delta - g^\delta 1 \tag{31}$$

$$\pi(i) = \delta(i) \quad \text{whenever} \quad t_i = 0 \tag{32}$$

Then $g^\pi \geq g^\delta$. Moreover, $g^\pi = g^\delta$ if and only if $s^\pi t = 0$, $s^\pi = s^\delta$ and $w^\pi \geq w^\delta + t$.

PROOF: The proof consists of premultiplying (31) by s^π and Y^π and examining the consequences. First premultiply (31) by s^π .

$$0 \leq s^\pi t = s^\pi R^\pi + 0 - g^\delta$$

$$0 \leq s^\pi t = g^\pi - g^\delta$$

Consequently, $g^\pi \geq g^\delta$, with equality holding if and only if $s^\pi t = 0$.

†We use "<" in such a way that $0 < t$ means $0 \leq t$ and $0 \neq t$.

For the remainder of the proof, suppose $s^\pi t = 0$. Expression (32) then implies that decisions $\delta(i)$ and $\pi(i)$ can only differ when $s_i^\pi = 0$. A direct consequence of this is

$$s^\pi P^\pi = s^\pi P^\delta . \tag{33}$$

Theorem 1 assures $s^\pi = s^\pi P^\pi$. This and (33) imply $s^\pi = s^\pi P^\delta$, so that a second application of Theorem 1 yields $s^\pi = s^\delta$. Consequently, $(P^\pi)^* = (P^\delta)^* = 1\, s^\delta$. This and (17) imply

$$Y^\pi(P^\pi - I) = (P^\pi)^* - I = 1s^\delta - I .$$

Now, premultiply (31) by Y^π and use the above to obtain

$$Y^\pi t = Y^\pi R^\pi + (1s^\delta - I)z^\delta - 0$$

$$Y^\pi t = w^\pi - w^\delta, \tag{34}$$

the last from Theorem 4. Next observe from (13) that

$$Y^\pi t = \sum_{i=0}^{\infty} [(P^\pi)^i - 1s^\pi]t$$

$$= \sum_{i=0}^{\infty} (P^\pi)^i t \geq t . \tag{35}$$

Combined, (34) and (35) form $t \leq w^\pi - w^\delta$, which completes the proof. ■

As applied to policy iteration, Lemma 1 guarantees that the gain-rate increases whenever a change occurs in a state that is recurrent (has $s_i > 0$) under the revised policy; if changes only occur in the transient states, then the bias increases, as $w^\pi \geq w^\delta + t > w^\delta$. This precludes cycling and thereby assures termination in finitely many repetitions of the policy improvement step.

THEOREM 5. The policy iteration routine terminates in finitely many iterations. The last policy γ evaluated is gain-optimal.

PROOF: Finite termination is already established. Let γ be the last policy evaluated. The policy improvement step is such that every policy δ satisfies

$$0 \geq R^{\delta} + (P^{\delta} - I)\, z^{\gamma} - g^{\gamma} 1.$$

Premultiply by s^{δ}.

$$0 \geq g^{\delta} - g^{\gamma} .$$

Hence, $g^{\gamma} \geq g^{\delta}$ for all δ, so that γ is gain optimal. ■

Remark: The fabric of Lemma 1 leads one to suspect that the policy iteration routine ends with a bias-optimal policy, not just a gain-optimal policy. This is not necessarily the case, as Example 2 will attest.

Linear Programming:

We shall establish an intimate connection between policy iteration and the linear program displayed below.

PROGRAM I: Minimize g

subject to the constraints

$$g + z_i - \sum_j P_{ij}^k z_j \geq R_i^k \qquad \text{all } i, k .$$

$$g, z_i \text{ unrestricted.}$$

THEOREM 6 . Program I has g^* as the optimal value of its objective function.

PROOF: Policy iteration terminates with a policy γ having $g^{\gamma} = g^*$ and having a solution (g^*, z^{γ}) of the policy evaluation equation such that every t_i^k is non-positive. Compare the equations defining t_i^k with the constraints of Program I to see that (g^*, z^δ) is a feasible solution to this program. Consider any feasible solution (g, z) to Program I. It suffices to show that $g \geq g^*$. Those constraints in Program I that correspond to decisions for policy γ combine to form the vector inequality

$$g1 + z - P^\gamma z \geq R^\gamma .$$

Premultiply the above by s^γ. This results in $g \geq s^\gamma R^\gamma = g^*$, which completes the proof. ∎

Observe that Theorem 6 fails to specify a procedure for recovering a gain-optimal policy from the final tableau of Program I. Analogy with the discounted model suggests picking for each i a decision k whose inequality in the final tableau of Program I is tight. However, in the undiscounted case, some state i can have every inequality slack. The difficulty is illuminated by considering the dual to Program I, which is itself extremely interesting. A slightly tampered dual to Program I is

PROGRAM II. Maximize $\sum_i \sum_k x_i^k R_i^k$

subject to the constraints

$$\sum_i \sum_k x_i^k = 1$$

$$(36) \qquad \sum_k x_i^k - \sum_j \sum_k x_j^k P_{ji}^k = 0 \qquad i = 2, \ldots, N$$

$$x_i^k \geq 0 \qquad \text{all } i, k.$$

Remarks: Formally, the dual includes constraints (36) for $i = 1, 2, \ldots, N$. However, the sum of these constraints is zero, and any one of them can be deleted. Of course, Program II has fewer rows than Program I.

Relationship of Linear Programming to Policy Iteration:

In the discounted Markov decision model, the feasible bases are in one-to-one correspondence with the policies. The situation is more complex here, as Example 1 will attest by exhibiting a feasible basis to Program II that corresponds to no policy.

The other half of the analogy with the discounted model is preserved, as we now demonstrate. Let δ be any policy, and set $x_i^k = 0$ whenever $k \neq \delta(i)$. The matrix B^δ is obtained from $(I - P^\delta)$ by replacing its first column by 1. The non-zero variables of Program II can be arrayed into the $1 \times N$ vector x, and the constraints require

$$xB^\delta = [1, 0, \ldots, 0] \qquad x \geq 0 .$$

Compare the above with (20). Theorem 3 assures $x = g^\delta$ and that B^δ, which is invertible, must be a basis. Moreover, the value of the objective function that is associated with this basis is easily seen to be g^δ.

Every policy corresponds in this way to a basic feasible solution to Program II. One might then hope that if Program II is initiated with a basis corresponding to a policy it executes a series of pivot steps, with each successive basis corresponding to a policy. To see what happens, we first rewrite the constraints as row vectors. (The simplex routine is normally described in terms of column vectors, but row vectors meld better with our scheme of notation.) Define the 1 by N vector

$$A_i^k = (1, -P_{i2}^k, \ldots, 1 - P_{ii}^k, \ldots, -P_{iN}^k) .$$

This allows the constraints of Program II to be rewritten as $x_i^k \geq 0$ and

$$\sum_i \sum_k x_i^k A_i^k = (1, 0, \ldots, 0).$$

With B^δ as the current basis, the row chosen to "enter" the basis is the one for which

$$R_i^k - A_i^k (B^\delta)^{-1} R^\delta$$

is the most positive. (This is the row-notation analogue of $c_j - c^B B^{-1} A_j$.) Equation (29) interprets $(B^\delta)^{-1} R^\delta$, in terms of g^δ and z^δ.

$$(37) \qquad R_i^k - A_i^k (B^\delta)^{-1} R^\delta = R_i^k - g^\delta - z_i^\delta + \sum_j P_{ij}^k z_j^\delta$$

Note that the right-hand side of (37) is precisely t_i^k, as defined in the policy improvement routine. This means that the simplex algorithm pivots in the row for which t_i^k is most positive. This is the result suggested by analogy with the discounted case. However, Example 1 illustrates that when A_i^k enters the basis, the simplex algorithm does not necessarily call for removal of row $A_i^{\delta(i)}$. But this is sensible: no calculation need be performed to make this decision; a basic feasible solution results; and Lemma 1 precludes cycling. This discussion is recapitulated in a modified pivot rule followed by Theorem 7, whose proof is now complete.

<u>Modified Exit Rule:</u> Suppose B^δ is the current basis. If the simplex algorithm calls for entry of row A_i^k, remove row $A_i^{\delta(i)}$.

THEOREM 7. The following procedures make the same sequence of pivots.

(i) The simplex routine, initiated with basis B^δ and applied to Program II, using the modified exit rule.
(ii) The policy iteration routine, initiated with policy δ, and, at each policy improvement step, changing only the one decision for which t_i^k is most positive.

<u>Remark:</u> Program II involves variables $\{x_i^k\}$, while policy iteration involves the dual variables $\{g_i, v_i\}$. It is widely known (cf. Wagner [16, pg. 147]) that application of the simplex method to the primal program is equivalent, pivot for pivot, to application of the dual simplex method to the dual program. Our altered exit rule for Program II corresponds

(in a way that is obvious to linear programming buffs) to an altered entry rule for application of the dual simplex method to Program I. As a consequence of Theorem 7, policy iteration with one variable changed per iteration is exactly the same as applying the dual simplex method to Program I, with the altered entry rule.

It has been claimed that: (1) Program I can terminate with all inequalities corresponding to a particular state slack; (2) Program II can have basic feasible solutions that fail to correspond to policies; and (3) the Program II sometimes pivots from a basic feasible solution corresponding to a policy to one that does not. Degeneracy accounts for each of these peculiarities, and Example 1 illustrates all three. It initiates Program II with a basis corresponding to a policy, pivots to a basis that does not correspond to a policy, and then terminates with a solution whose dual has the lone inequality corresponding to state 2 slack.

Example 1: There are three states, two of which are transient, and only two policies. Policy iteration terminates with the policy γ having $\gamma(3) = 2$. If linear programming is initiated with the basis corresponding to the other policy, it pivots in row A_3^2, (as expected) removes row A_2^1, and then terminates. So, the final basis has two rows for state 3 and none for state 2. Moreover, the lone inequality in Program I corresponding to state 2 holds strictly. The calculations necessary to verify this are left to the reader. ∎

State	Decision	Transition Probabilities			Rewards
i	k	P_{i1}^k	P_{i2}^k	P_{i3}^k	R_i^k
1	1	1	0	0	0
2	1	1/3	1/3	1/3	-1
3	1	1/3	1/3	1/3	-1
3	2	1	0	0	0

Table 1: An example in which the simplex algorithm differs from policy iteration.

Incidentally, the natural interpretation of the variable x_i^k in Program II is as the joint probability of observing state i and selecting decision k . This allows one to argue, as Wagner [15] did, that a stationary non-randomized policy is optimal over the class of all stationary randomized policies.

<u>A Bias-Optimal Policy</u>: We are now prepared to show that a bias-optimal policy exists, that the policy iteration routine depicted in Figure 2 may not find one, but that a simple adaptation of it does so. First note from Theorem 5 that policy iteration terminates by identifying a policy γ having $t_i^k \leq 0$ for all i and k and $t_i^k = 0$ whenever $k = \gamma(i)$. Policy γ has gain rate $g^\gamma = g^*$ and a solution (any solution will do) (g^γ, z^γ) of the policy evaluation equation. Recall that D_i is the set of decisions for state i, that Δ is the set of policies, and set

$$D_i^* = \{k \in D_i \mid t_i^k = 0\}$$

$$\Delta^* = \{\delta \in \Delta \mid \delta(i) \in D_i^* \quad \text{for } i = 1, \ldots, N\} .$$

<u>Remark:</u> Adding a scalar to each element of z^γ does not change any t_i^k. So, although z^γ is ambiguously defined, D_i^* is not. Of course, γ is contained in Δ^*, but Δ^* contains additional policies when any state i has multiple t_i^k equal to zero.

We shall soon show that a bias-optimal policy exists and is contained in Δ^*. Note that Δ^* is the set of those policies δ such that

$$(38) \qquad 0 = R^\delta + (P^\delta - I)\, z^\gamma - g^* 1 .$$

This is the policy evaluation equation; so Theorem 4 assures $g^\delta = g^*$ and

$$(39) \qquad w^\delta = z^\gamma - (s^\delta z^\gamma) . \qquad \delta \in \Delta^* .$$

Consequently, maximizing w^δ over Δ^* is equivalent to maximizing the scalar $s^\delta(-z^\gamma)$ over Δ^*. The latter is the gain rate for policy δ in the following <u>altered</u> Markov decision process:

(1) the states are $1, 2, \ldots, N$

(2) for each state, the set of alternative decisions is D_i^*

(3) the immediate reward for observing state i is z_i^γ (independent of k)

(4) the transition probabilities are P_{ij}^k, as before.

Remark: The immediate return $-z_i^\gamma$ is data in the altered problem, but it originally depended on the policy and, even for a fixed policy, contained one degree of freedom.

Clearly, the techniques of policy iteration and linear programming can be applied to maximize the gain rate $s^\delta(-z^\gamma)$ for the altered problem. Theorem 8 shows, in effect, that the policies in Δ^* dominate those that are not in Δ^*.

THEOREM 8. Let γ denote the last policy evaluated by the policy iteration routine in Figure 2. Let policy λ maximize $s^\lambda(-z^\gamma)$ over Δ^*. Then, policy λ is bias-optimal.

PROOF: The variant of Lemma 1 in which all inequalities are reversed is also true, and the proof is a direct application of this variant. Let π be any policy that is not contained in Δ^*. Policy iteration terminates with

$$(40) \qquad 0 > t^\pi = R^\pi + (P^\pi - I)z^\gamma - g^* 1 .$$

Pick policy δ such that, for each i, decision $\delta(i)$ is contained in D_i^* and, in addition

$$(41) \qquad \delta(i) = \pi(i) \text{ whenever } t_i^\pi = 0 .$$

So, policy δ is in Δ^* and the variables (g^*, z^γ) satisfy its policy evaluation equation, as is seen in (38). Compare (40)-(41) with (31)-(32) to see that π and δ satisfy the hypothesis of Lemma 1, with the inequality reversed. As a consequence, either $g^\pi < g^\delta = g^*$ or $g^\pi = g^*$ and $w^\pi < w^\delta$.

Since $w^\delta \le w^\lambda$, we have shown that λ is bias-optimal. ■

The last line of the proof of Theorem 8 supports the following corollary.

COROLLARY 2: Δ^* contains all bias-optimal policies.

So, if Δ^* contains a single policy, that policy must be bias-optimal. If Δ^* contains several policies, the policy iteration routine in Figure 2 will not distinguish between them. Showing that the original policy iteration routine can stop short of a bias-optimal policy amounts to finding a model in which there are ties (i.e., multiple t_i^k equal to zero) amongst the recurrent states at termination of policy termination.

Example 2: Consider the model depicted in Table 2, which has two options for state 1 and none for states 2 and 3. Income is earned only when escape from state 1 occurs. One can readily calculate (or verify from renewal theory) that selecting either option for state 1 results in a gain-rate of 1. Moreover, when policy iteration is initiated with either option, all t_i^k equal zero, and termination is immediate. The policy λ with $\lambda(1) = 1$ is bias-optimal. So, to stop short of this, initiate policy iteration with $\delta(1) = 2$. ■

State	Decision	Transition Probabilities			Rewards
i	k	P_{i1}^k	P_{i2}^k	P_{i3}^k	R_i^k
1	1	1/2	1/2	0	0
1	2	2/3	0	1/3	0
2	1	1	0	0	3
3	1	1	0	0	4

Table 2. An example in which policy iteration stops short of a bias-optimal policy.

Long Planning Horizons

This section treats the undiscounted Markov decision model when the planning horizon is long, but fixed and finite. Let $f(n)$ be the $N \times 1$ vector whose ith component is the maximum total expected income when the initial state is i and when the planning horizon is n epochs. With $f(0) = 0$, the standard recursive relation of dynamic programming characterizes $f(n)$ by the equation system

$$(42) \qquad f(n)_i = \max_k \{R_i^k + \sum_j P_{ij}^k f(n-1)_j\} \qquad n \geq 1 .$$

Of course, (42) allows computation of $f(n)$ in increasing n, as only $f(n-1)$ appears on its right-hand side. It is intuitively clear (and easily proven) that $f(n)$ is attained by a decision procedure

$$\Pi = (\ldots, \pi^n, \ldots, \pi^2, \pi^1) ,$$

where, for each n, π^n is a non-randomized policy satisfying

$$(43) \qquad f(n) = R^{\pi^n} + P^{\pi^n} f(n-1) \qquad n \geq 1 .$$

A decision procedure Π attaining (43) for each n is called time-optimal. For the remainder of this section we reserve the symbol Π to denote a time-optimal decision procedure. Writing Π in this way highlights the fact that π^n is the policy used when n is the number of epochs to the end of the planning horizon.

Two Counterexamples:

This section analyzes the asymptotic behavior of $f(n)$ and π^n as n approaches infinity. Theorems describing such behavior are often called turnpike theorems. As motivation, we first provide counterexamples to analogues of two turnpike theorems for the discounted model. In the discounted model, a stationary non-randomized policy comes arbitrarily close to being optimal for the finite-horizon problem, providing the horizon is long enough. The reason this is the case is that, as the planning horizon grows long, whatever income might be acrued from recognizing the end as it approaches is discounted by a factor that approaches zero. Example 3 verifies that this turnpike result does not carry over to the undiscounted case, where the extra income maintains full value, rather than being discounted.

Example 3. The model depicted in Table 3 has two states and two policies. Selecting decision 2 for state 1 earns \$1 this period and incurs a cost of \$10 next period
bility .5 . So decision 2 is inadvisable, except at the final epoch. Consequently, $\pi^1(1) = 2$ and $\pi^n(i) = 1$ for all other combinations, which lead to

$$f(n) = \begin{pmatrix} 1 \\ -9 \end{pmatrix} \qquad n = 2, 3, \ldots \quad .$$

Moreover, the bias-optimal policy λ has $\lambda(1) = 1$, which implies $g^* = 0$ and $w^* = (0, -10)$. Consequently,

$$f(n) - ng^* 1 - w^* = \begin{pmatrix} 1 \\ 1 \end{pmatrix} , \quad n = 2, 3, \ldots \quad . \tag{44}$$ ■

State i	Decision k	Transition Probabilities P^k_{i1}	P^k_{i2}	Reward R^k_i
1	1	1	0	0
1	2	.5	.5	1
2	1	1	0	-10

Table 3. As the end approaches ...

A decision procedure $\Delta = (\ldots, \delta^2, \delta^1)$ is called eventually stationary if some m exists such that $\delta^n = \delta^m$ whenever $n \geq m$. Is the time-optimal policy eventually stationary? One might hope so, since the "effect" of the end of the planning horizon on decisions made at the beginning can be expected to vanish as the planning horizon approaches infinity. But this need not be the case, and the difficulty is best illustrated by example.

In preparation, let us compute the gain and bias for the discrete-time Markov decision process whose transition matrix $P(x)$ and reward vector $R(x)$ are given below as functions of the parameter x .

$$P(x) = \begin{pmatrix} x & 1-x \\ 1 & 0 \end{pmatrix} \quad R(x) = \begin{pmatrix} 2-x \\ 0 \end{pmatrix} \tag{45}$$

The steady-state probability $s(x)_1$ of observing state 1 satisfies $s(x)_1 = x\, s(x)_1 + 1 - s(x)_1$; hence

$$s(x)_1 = 1/(2 - x) . \tag{46}$$

Consequently, the gain-rate $g(x)$ is 1 , independent of x. With $z_1 = 0$, the solution $(1, z)$ of the policy evaluation equation has $z_2 = -1$. Theorem 4 than assures

$$w(x) = \begin{pmatrix} 1 - s(x)_1 \\ - s(x)_1 \end{pmatrix} . \tag{47}$$

Note from (46) and (47) that decreasing x decreases $s(x)_1$ and thereby increases $w(x)$. Example 4 will offer a choice of x . We have just seen that the bias-optimal policy will be the one with the smaller x .

<u>Example 4.</u> The model depicted in Table 4 has two states and two policies. Fix scalars a and b with $0 < a < b < 1$. As noted above, the bias-optimal policy λ has $\lambda(1) = a$. We shall see that the time-optimal policy Π oscillates between a and b . Proof that this occurs will amount to observing that $T(n)$ oscillates, where $T(n) = f(n)_1 - f(n)_2 - 1$. Equation (42) states that $f(n)_1$ is attained by the larger of the two expressions:

$$2 - a + a\, f(n-1)_1 + (1-a)\, f(n-1)_2$$

$$2 - b + b\, f(n-1)_1 + (1-b)\, f(n-1)_2 \quad .$$

State i	Decision k	Transition Probability P^k_{i1}	P^k_{i2}	Reward R^k_i
1	a	a	1-a	2-a
1	b	b	1-b	2-b
2	1	1	0	0

Table 4. A cyclic policy is optimal.

Subtract the second expression from the first; the result is $(a - b)\ T(n - 1)$. This means that $\pi^n(1) = a$ whenever $T(n - 1)$ is negative and that $\pi^n(1) = b$ whenever it is positive. A routine calculation shows that if $T(n - 1)$ is negative, then $T(n) = (a - 1)\ T(n - 1)$, which is positive. Similarly, if $T(n - 1)$ is positive, then $T(n) = (b - 1)T(n - 1)$, which is negative. Since $T(0) = -1$, the sequence $\{T(n)\}$ oscillates in sign, and π^n equals a and b when n is odd and even, respectively.

To compute the asymptotic form of $f(n)$, one can exploit the fact that $\pi^n(1)$ oscillates.† Working first with n even, reward vector $R(b) + P(b)\ R(a)$ and transition matrix $P(b)\ P(a)$, one can calculate the gain and bias from Theorem 4. There results

$$f(n) = n\begin{pmatrix}1\\1\end{pmatrix} + \begin{pmatrix}1 - s_1\\ -s_1\end{pmatrix} + o(1) \qquad \text{as } n\to\infty, \tag{48}$$

with $s_1 = a/(a + b - ab)$. The bias-optimal policy λ has $g^* = 1$ and w^* given by (47), with $x = a$. So

†The recursive relation for $T(n)$ also yields an exact, but uninteresting formula for $f(n)$.

$$f(n) - ng^{*}1 - w^{*} = d1 + o(1) , \tag{49}$$

where $d = 1/(2-a) - a/(a+b-ab)$. A little algebra verifies that d is positive. ∎

Examples 3 and 4 exhibit several interesting features. Compare (44) with (49) to see that both examples give rise to a scalar d, independent of the state, which is the asymptotic advantage of recognizing the end of the planning horizon as it approaches. This is a turnpike result, and Theorem 9 will show that it holds in general. The following plausibility argument may make this result seem more intuitive. Suppose decision-making has gone on for a long time but that we are still nowhere near the end of the planning horizon. As the end is nowhere near, the extra income earned for recognizing it as it approaches has yet to be earned. On the other hand, we are so far from the beginning of the planning horizon that the current state yields negligible information as to what state the process started in. Consequently, the extra income would seem to be independent of the starting state.

Note from (49) that little is lost if one uses the time-optimal policy for the last epochs and a bias-optimal policy before then. This property also holds in general and appears as a corollary to Theorem 9. Example 4 exhibits a particularly curious feature. Suppose we increase b. Decision a remains bias-optimal, as the bias for decision b decreases. However, the number s_1 in (48) also decreases, so that it becomes more desirable to alternate decisions a and b, though less desirable to use decision b exclusively.

A Turnpike Theorem:

The turnpike result in (49) is obtained from analyzing the asymptotic behavior of

$$e(n) = f(n) - ng^{*}1 - w^{*} \qquad n = 0, 1, \ldots \quad .$$

Note that $e(0) = -w^{*}$ and recall that $s^{\lambda}w^{*} = 0$. This means that $e(0)$ normally has positive and negative elements. Theorem 9 will establish a non-negative scalar d such that $e(n) \to d1$ as $n \to \infty$. In preparation, we use (42) to establish an interesting recursive relation for $e(n)$:

$$e(n)_i = \max_k \{R_i^k + \sum_j P_{ij}^k f(n-1)_j - ng^* - w_i^*\}$$

$$= \max_k \{R_i^k + \sum_j P_{ij}^k [e(n-1)_j + (n-1)g^* + w_j^*] - ng^* - w_i^*\}$$

(50) $$= \max_k \{\bar{R}_i^k + \sum_j P_{ij}^k e(n-1)_j\} \qquad n = 1, 2, \ldots$$

where

$$\bar{R}_i^k = R_i^k + \sum_j P_{ij}^k w_j^* - w_i^* - g^* .$$

Clearly, (50) is the functional equation for the sequential decision process with reward $\bar{R}_i^k$ for selecting decision k when in state i and with terminal reward $-w_i^*$, which is $e(0)_i$. Compare $\bar{R}_i^k$ as just defined with the value of t_i^k at termination of policy iteration. They are identical! So, $\bar{R}_i^k$ is non-positive. It equals zero if and only if k is in D_i^*. In that some $\bar{R}_i^k$ are negative and others are zero, the model resembles a stopping problem.

LEMMA 2. Let λ be a bias-optimal policy, and let $\Pi = (\ldots, \pi^2, \pi^1)$ be a time-optimal decision procedure. Then

(51) $$P^\lambda e(n) \le e(n+1) \le P^{\pi^{n+1}} e(n) .$$

PROOF: By definition,

$$e(n+1) = \bar{R}^{\pi^{n+1}} + P^{\pi^{n+1}} e(n) \le P^{\pi^{n+1}} e(n)$$

since $\bar{R}^\pi \le 0$ for any π. Similarly, since $\bar{R}^\lambda = 0$,

$$e(n+1) \ge \bar{R}^\lambda + P^\lambda e(n) = P^\lambda e(n) . \qquad \blacksquare$$

Since P^δ is stochastic for each policy δ, it is a simple consequence of (51) that $\min_i e(n)_i$ is nondecreasing in n and that $\max_i e(n)_i$ is nondecreasing in n. In particular, since $e(0) = -w^*$,

$$\min_j(-w_j^*) \leq e(n)_i \leq \max_j(-w_j^*) \qquad \text{all } i, \quad n = 1, 2, \ldots . \tag{52}$$

THEOREM 9. For some nonnegative scalar d, $e(n) \to d1$ as $n \to \infty$.

PROOF: Our proof is somewhat involved. Set

$$m_i = \liminf_{n \to \infty} e(n)_i \qquad M_i = \limsup_{n \to \infty} e(n)_i$$

$$m = \min_i m_i \qquad M = \max_i M_i .$$

Inequality (52) assures that m and M are finite. Our proof consists of showing that m = M, and this is accomplished in stages. Consider

Proposition 1: $s_i^\lambda > 0 \Rightarrow M_i = m_i = m$. To obtain (53) and (54), use (51) repeatedly and then let $r \to \infty$.

$$e(n + r) \geq (P^\lambda)^r e(n) \tag{53}$$

$$m_i \geq s^\lambda e(n) \qquad \text{all } i, \text{ all } n . \tag{54}$$

Fix i, temporarily, and pick a sequence $\{n_\ell\}$ such that $e(n_\ell)_i \to M_i$ and $\ell \to \infty$. Take limits in (54).

$$\begin{aligned} m_i &\geq \liminf_{\ell \to \infty} \{s^\lambda e(n_\ell)\} \\ &= s_i^\lambda M_i + \liminf_{\ell \to \infty} \sum_{j \neq i} \{s_j^\lambda e(n_\ell)_j\} \\ &\geq s_i^\lambda M_i + \liminf_{n \to \infty} \sum_{j \neq i} \{s_j^\lambda e(n)_j\} , \end{aligned}$$

the last by set inclusion. Note that the lim inf is superadditive; i.e., the lim inf of a sum is no smaller than the sum of the lim inf's. Consequently,

(55) $$m_i \geq s_i^\lambda M_i + \sum_{j \neq i} s_j^\lambda m_j \qquad \text{all } i .$$

Since $M_i \geq m_i$, minimizing (55) over i gives

(56) $$m \geq \sum_j s_j^\lambda m_j .$$

Equation (56) could have been obtained from a less delicate analysis; but we also rearrange (55) and introduce a slack variable t_i to obtain

(57) $$m_i = s_i^\lambda(M_i - m_i) + \sum_j s_j^\lambda m_j + t \qquad \text{all } i ,$$

where $t_i \geq 0$. Premultiply (57) by s_i^λ, sum over all i, and then cancel two terms.

(58) $$0 = \sum_i (s_i^\lambda)^2 (M_i - m_i) + \sum_i s_i^\lambda t_i .$$

The right hand side of (58) has non-negative multiplicands. Consequently whenever $s_i^\lambda > 0$, it must be that $m_i = M_i$ and that $t_i = 0$. This means that, in addition, when s_i^λ is positive, (57) reduces to

$$m_i = \sum_j s_j^\lambda m_j \leq m ,$$

the last from (56). This completes the proof of proposition 1.

Proposition 1 treats the set of states that are recurrent under policy λ. Let

$$A = \{i \mid s_i^\lambda > 0\} \qquad B = \{i \mid M_i = M\} .$$

It is easily seen that the set A is closed under P^λ. Proposition 2, which follows, will verify that B is closed under P^π, where π constructed by this procedure. Fix i, temporarily. Pick a decision $\pi(i)$ such that $\pi^{n_\ell+1}(i) = \pi(i)$ for an infinite sequence $\{n_\ell\}$ such that $e(n_\ell+1) \rightarrow M_i$. Repeat for each i, <u>including</u> alteration of the sequence $\{n_\ell\}$ as required.

Proposition 2: The set B is closed under P^π. Taking limits in (51) yields

$$M_i \leq \limsup_{\ell \to \infty} \left(\sum_j P^\pi \right)_{ij} e(n_\ell)_j \qquad \text{all } i$$

$$(59) \qquad \leq \sum_j (P^\pi)_{ij} M_j \qquad \text{all } i,$$

the last since "lim sup" is subadditive. For $i \in B$, one has $M_i = M$. Subtract M from both sides of (59). For i in B,

$$0 \leq \sum_j (P^\pi)_{ij}(M_j - M) \leq 0 .$$

Equality must hold in the above, which verifies Proposition 2.

Finally, suppose $m < M$. Proposition 1 verifies that $M_i = m$ whenever $i \in A$. So, A and B are disjoint. Define policy δ such that $\delta(i) = \lambda(i)$ for i in A and $\delta(j) = \pi(j)$ for j in B. Policy δ has two disjoint closed sets and, therefore, at least two ergodic chains. This contradicts our hypothesis and proves $m = M = d$. To see that d is nonnegative, simply note from (51) that

$$e(n) \geq (P^\lambda)^n e(0) = (P^\lambda)^n (-w^*) .$$

Since $s^\lambda w^* = 0$, taking limits in the above verifies $m_i \geq 0$ for each i, completing the proof of Theorem 9. ■

Remark: The above proof of Theorem 9 is nonconstructive. In particular, it does not specify a finite procedure for calculating d. No such procedure is now known to the author.

It was observed earlier that $\bar{R}_i^k \leq 0$ with equality holding if and only if k in D_i^*. Recall from the section on policy iteration that Δ^* is the set of policies δ such that, for each i, $\delta(i)$ is in D_i^*.

COROLLARY 3: π^n is in Δ^* for all sufficiently large n.

PROOF: Let $-\epsilon = \max\{\bar{R}_i^k \mid \bar{R}_i^k < 0\}$. Pick N big enough that $\|e(n) - d1\| < \epsilon/2$ for $n \geq N$. For $n > N$ and k not in D_i^*,

$$\bar{R}_i^k + \sum_j P_{ij}^k e(n-1)_j < -\epsilon + \epsilon/2 + d < e(n).$$

Consequently $\pi^n(i)$ cannot equal k. ■

A simple implication of this corollary is that Π is eventually stationary whenever Δ^* contains a single policy, which is usually the case. The vector $V(m,n)$ is defined in terms of a planning horizon that is $m + n$ epochs long. Let the ith element of $V(m,n)$ be the expected income for starting in state i, using the bias-optimal policy λ for the first m epochs and using the time-optimal policy Π for the remaining epochs. Combine (24) and the definition of $e(n)$ to obtain

$$V(m,n) = mg^*1 + w^* - (P^\lambda)^m w^* + (P^\lambda)^m\{ng^*1 + w^* + e(n)\}$$

Two terms cancel.

$$V(m,n) = (m+n)g^*1 + w^* + (P^\lambda)^m e(n) \quad (60)$$

COROLLARY 4: For any positive ϵ, there exists an N such that for every $n \geq N$ and every m,

$$f(m+n) - V(m,n) \leq \epsilon 1 \quad .$$

PROOF: The corollary follows immediately from (60) and $e(n) \rightarrow d1$. ■

So, we can use the bias-optimal policy until the end of the planning horizon approaches and obtain within ϵ of the optimum.

Bibliographic Notes

The Markov decision model has a considerable history, even prior to Howard's excellent book [7], which kindled intense interest in this model. Of the many papers it inspired, only the most directly relevant are mentioned here. Using perturbation theory, Schweitzer [11, 12] first showed that P^* and Z can be computed with a single inversion, but (23) is new. The policy evaluation equation, (28), was interpreted by Howard [7] in terms of the gain-rate and relative values. Blackwell [1] observed the relevance of the fundamental matrix and demonstrated (30), which removed the ambiguity in the relative values. Theorem 4 itself is due to Denardo [4], following Veinott [13]. Howard [7] suggested policy iteration, and Blackwell [1] showed that it works for the model considered here. Manne [10] and, independently, DeGhellinck [3] introduced the linear programming formulation in Program II. The close tie between linear programming and policy iteration has long been known, but the revised exit rule is new. Veinott [13] first showed how to compute a bias-optimal policy by dynamic programming, though Theorem 8 follows Denardo [4]. Long planning horizons had earlier been considered by Brown [2], Schweitzer [11], and Lanery [9], and the development here borrows from all three. Example 4 is essentially due to Brown [2]. Schweitzer [11] provided the interpretation of e(n) in (50).

REFERENCES

1. Blackwell, D., "Discrete Dynamic Programming," Ann. Math. Statist., Vol. 33, 719-726, 1962.

2. Brown, B. W., "On the Iterative Method of Dynamic Programming on a Finite Space Discrete Time Markov Process," Ann. Math. Statist., Vol. 36, 1279-1285, 1965.

3. DeGhellinck, G., "Les Problèmes De Décisions Séquentielles," Cahiers Centre Études Recherche Opér. 2, 161-179, 1960.

4. Denardo, E. V., "Computing a Bias-optimal Policy in a Discrete-time Markov Decision Problem," Operations Research, Vol. 18, 279-289, 1970.

5. Denardo, E. V., Elements of Dynamic Programming, (in preparation).

6. Feller, W., An Introduction to Probability Theory and Its Applications, Vol. I., Wiley, New York, 1950.

7. Howard, R. A., Dynamic Programming and Markov Processes, Wiley, New York, 1960.

8. Kemeny, J. G., and J. L. Snell, Finite Markov Chains, Van Nostrand, Princeton, 1960.

9. Lanery, E. "Étude Asymptotique des Systèmes Markoviens à Commande," R.I.R.O., Vol. 1, 3-56, 1967.

10. Manne, A., "Linear Programming and Sequential Decisions," Management Science, Vol. 6, 259-276, 1960.

11. Schweitzer, P. J., Perturbation Theory and Markovian Decision Processes, Ph. D. Thesis, M.I.T., 1965.

12. Schweitzer, P. J., "Perturbation Theory and Finite Markov Chains," J. Appl. Prob., Vol. 5, 403-413, 1968.

13. Veinott, A. F., Jr. "On Finding Optimal Policies in Discrete Dynamic Programming with no Discounting," Ann. Math. Statist., Vol. 37, 1284-1294, 1966.

14. Wagner, H. M., "On the Optimality of Pure Strategies," Management Science, Vol. 6, 268-269, 1969.

15. Wagner, H. M., Principles of Operations Research with Applications to Managerial Decisions, Prentice Hall, 1969.

Department of Administrative Sciences
Yale University
New Haven, Connecticut 06520

This research was supported by National Science Foundation grant GK-13757.

On the Perfect Graph Theorem

D. R. FULKERSON

In a recent brilliant paper [7] L. Lovász has settled in the affirmative a conjecture due to Berge [1, 2] that had been outstanding in graph theory for over a decade, the perfect graph conjecture. In his paper Lovász gives two separate proofs of the conjecture. The first proof uses combinatorial results I had obtained earlier about anti-blocking pairs of polyhedra [3, 4, 5]. Concerning this proof, Lovász states: "It should be pointed out that thus the proof consists of two steps, and the most difficult second step was done first by Fulkerson." I would be less than candid if I did not say that I agree with this remark, at least in retrospect. But the fact remains that, while part of my aim in developing the anti-blocking theory had been to settle the perfect graph conjecture, and that while I had succeeded via this theory in reducing the conjecture to a simple lemma about graphs [3, 4] (the "replication lemma", a proof of which is given in this paper) and had developed other seemingly more complicated equivalent versions of the conjecture [3, 4, 5], I eventually began to feel that the conjecture was probably false and thus spent several fruitless months trying to construct a counterexample. It is not altogether clear to me now just why I felt the conjecture was false, but I think it was due mainly to one equivalent version I had found [4, 5], a version that does not explicitly mention graphs at all.

Theorem 1. Let A be a $(0,1)$-matrix such that the linear program $yA \geq w$, $y \geq 0$, $\min 1 \cdot y$ (where $1 = (1,\ldots,1)$) always has an integer solution vector y whenever w is a $(0,1)$-vector. Then this program always has an integer solution vector y whenever w is a nonnegative integer vector.

While this version of the perfect graph theorem says nothing explicitly about graphs, it should be noted that a $(0,1)$-matrix A which satisfies the hypothesis of Theorem 1 must be the incidence matrix of maximal cliques† vs. vertices of a graph G, and hence the hypothesis can be rephrased to say: For any vertex-generated subgraph H of G (the vertices of H correspond to components of w that are 1), one can cover all vertices of H by assigning integer (and hence $(0,1)$-valued) weights (components of y) to cliques of H just as efficiently as one can by allowing fractional weights on cliques of H, efficiency being measured in terms of the component-sum of y. The conclusion then asserts that under these conditions, if one assigns vertices of G arbitrary nonnegative integer weights (components of w), the same phenomenon results: the rational covering program $yA \geq w$, $y \geq 0$, $\min 1 \cdot y$, always has an integer solution.

My proof of the equivalence of this assertion and the perfect graph conjecture was based on certain facts about anti-blocking pairs of polyhedra and on linear-programming duality arguments. This proof has not been published. But Lovász has given a combinatorial proof of Theorem 1 in [7], using the perfect graph theorem as one main tool.

To describe the perfect graph theorem, we begin by defining some well-known integer-valued functions of an arbitrary graph. Let G be a graph and let $\gamma(G)$, $\lambda(G)$, $\pi(G)$, and $\omega(G)$ denote respectively the chromatic number of G (the minimum number of anti-cliques‡ (independent sets of

---

†A clique in graph G is a subset of vertices of G, each pair of which is joined by an edge in G.

‡An anti-clique in graph G is a subset of vertices of G, no pair of which is joined by an edge, i.e. an anti-clique is a clique in the complementary graph $\bar{G}$ of G.

vertices) that cover all vertices of G), the clique number of G (the size of a largest clique in G), the partition number of G (the minimum number of cliques that cover all vertices of G), and the anti-clique number (internal stability number) of G (the size of a largest anti-clique in G). Clearly one always has the inequalities $\gamma(G) \geq \lambda(G)$, $\pi(G) \geq \omega(G)$. The graph G is γ-perfect if $\gamma(H) = \lambda(H)$ for every vertex-generated subgraph H of G ; G is π-perfect if $\pi(H) = \omega(H)$ for every vertex-generated subgraph H of G; G is perfect if it is both γ-perfect and π-perfect, i.e. if both G and its complementary graph $\bar{G}$ are γ-perfect (or π-perfect).

Theorem 2 (Perfect graph theorem). *If* G *is* γ-*perfect* (*or* π-*perfect*), *then* G *is perfect.*

A stronger form, one that is still open, asserts that G is perfect if and only if neither G nor its complement $\bar{G}$ contains an "odd hole" (an odd chordless circuit of size ≥ 5). Recently A. C. Tucker has proved this for planar graphs G [9]; planarity is a severe restriction, however, in studying perfection in graphs.

There are numerous classes of graphs that are known to be perfect. Some of the better-known examples are interval graphs, rigid-circuit graphs, comparability graphs, and in particular, bipartite graphs.

Now let A and B be the m by n and r by n incidence matrices of the families of maximal cliques and maximal anti-cliques, respectively, of G , where A has rows $a^1, \ldots, a^m$, B has rows $b^1, \ldots, b^r$. Define functions $\gamma_G(w)$, $\lambda_G(w)$, $\pi_G(w)$, $\omega_G(w)$, where w is a nonnegative integer n-vector, as follows. Let $\gamma_G(w)$ be the minimum in the integer linear program $yB \geq w$, $y \geq 0$, min $1 \cdot y$; let $\pi_G(w)$ be the minimum in the integer linear program $yA \geq w$, $y \geq 0$, min $1 \cdot y$; let $\lambda_G(w) = \max_{1 \leq i \leq m} a^i \cdot w$; let $\omega_G(w) = \max_{1 \leq j \leq r} b^j \cdot w$. Again one has the inequalities $\gamma_G(w) \geq \lambda_G(w)$, $\pi_G(w) \geq \omega_G(w)$. Say that G is γ-pluperfect if $\gamma_G(w) = \lambda_G(w)$ for all w , that G is π-pluperfect if $\pi_G(w) = \omega_G(w)$

for all w, and that G is pluperfect if it is both γ-pluperfect and π-pluperfect. (Note that γ-perfection, for example, would require only $\gamma_G(w) = \lambda_G(w)$ for all (0,1)-vectors w, rather than for all nonnegative integer vectors w.) One of the main combinatorial consequences of the theory of anti-blocking pairs of polyhedra is [3,4,5]:

Theorem 3(Pluperfect graph theorem). *If G is γ-pluperfect (or π-pluperfect), then G is pluperfect.*

Thus to prove the perfect graph theorem, it would suffice to show that if G is π-perfect, say, then G is also π-pluperfect. For this it would suffice to prove the following simple lemma [3,4].

Lemma (Replication lemma). *If G is π-perfect, and if we duplicate an arbitrary vertex in G, the resulting graph G' is π-perfect.*

Proof of replication lemma. Suppose G is π-perfect, and let G' be formed from G by duplicating vertex v, i.e. adjoin v' to G and join v' to all neighbors of v in G. It is enough to show that $\pi(G') = \omega(G')$. If v belongs to a maximum cardinality anti-clique in G, this is trivial, since $\omega(G') = \omega(G)+1$. Suppose on the other hand that v is in no maximum anti-clique of G. We show in this case that there is a minimum cover of G by cliques in which v is doubly covered. Let $C_1, \ldots, C_k$ be a minimum cover of G by cliques with $v \in C_1$. Suppress $C_1 - v$ (which is nonempty) to obtain graph H. Now $\omega(H) = k-1$, since $C_2, \ldots, C_k$ cover $H-v$ and v is in no maximum anti-clique of G. Since H is π-perfect, we can cover H by cliques $K_2, \ldots, K_k$ of H, with $v \in K_2$, say. These are cliques in G, and hence G is covered by $C_1, K_2, \ldots, K_k$, and $v \in C_1$, $v \in K_2$. This proves the assertion, whence it follows that $\pi(G') = \omega(G')$ in this case as well, proving the lemma.

One of Lovász's two proofs of the perfect graph theorem uses the pluperfect graph theorem and a stronger version of the replication lemma: The stronger version asserts that replacing a vertex v of a π-perfect graph G by a π-perfect graph H yields a π-perfect graph G'. His other proof, which is self-contained and elegant, also uses his replacement lemma. While this proof does not explicitly invoke the pluperfect graph theorem, it again brings out the central role played by the notion of pluperfection in studying perfection. He describes the theorem and proof in terms of hypergraphs [2] (hypergraph = family of sets), but the description in this form is only apparently more general, not actually so (just as Theorem 1 is more general in appearance only than a theorem about graphs). One can describe this proof, in purely graph-theoretic terms, as follows.

Proof of perfect graph theorem. Suppose that G is π-perfect, so that $\pi(H) = \omega(H)$ for every vertex-generated subgraph of G. Again it is sufficient to show that $\gamma(G) = \lambda(G)$. The proof can be viewed as an induction on $\lambda(G) = k$. The case $k = 1$ gives no difficulty. Consider the case $k > 1$. It is enough to find a maximal anti-clique D in G such that the graph G-D obtained from G by supressing vertices of D has $\lambda(G-D) = k-1$. Assume, on the contrary, that for every maximal anti-clique D in G, there is a clique C(D), no vertex of which is in D, of size k, i.e., assume that graph G-D has a clique C(D) of size k. Consider the weight vector w having components w(v), one for each vertex v of G, defined as follows: Let w(v) be the number of times v occurs in the family of sets C(D), as D ranges over all maximal anti-cliques of G. Let G' be the graph formed from G by replacing each vertex v in G by w(v) replicates of v. (If $w(v) = 0$, this means that v is suppressed in forming G'.) It follows from the replication lemma (or from Lovász's replacement lemma) that G' is π-perfect. If G has m maximal anti-cliques $D_1, D_2, \ldots, D_m$, then G' has mk vertices. Since the size of a maximum clique in G' is at most k, we have $\pi(G') \geq m$. On the other hand, a given anti-clique D_i of G contains at most one vertex of $C(D_j)$, and no vertex of $C(D_i)$, and hence

$$\omega(G') = \max_{1 \leq i \leq m} \sum_{v \in D_i} w(v)$$

$$= \max_{1 \leq i \leq m} \sum_{j=1}^{m} |D_i \cap C(D_j)|$$

$$\leq m - 1 ,$$

contradicting the fact that G' is π-perfect. This proves Theorem 2.

Observe that the contradiction obtained in the proof is gotten by a shrewd choice of the integer weight vector w.

There is another interesting characterization of pluperfect graphs (and hence of perfect graphs) that was given in [3, 4] and called there the max-max inequality. This characterization can be described as follows. Again let A and B be the m by n and r by n incidence matrices of the families of maximal cliques and anti-cliques of a graph G, where A has rows $a^1, \ldots, a^m$, one for each maximal clique, and B has rows $b^1, \ldots, b^r$, one for each maximal anti-clique. Let ℓ and w be nonnegative n-vectors, whose components correspond to vertices of G.

Theorem 4. *The graph G is pluperfect if and only if*

$$(*) \qquad (\max_{1 \leq i \leq m} a^i \cdot \ell)(\max_{1 \leq j \leq r} b^j \cdot w) \geq \ell \cdot w$$

for all nonnegative weight vectors ℓ and w.

In other words, the "weight" of a largest clique, computed using ℓ, times the "weight" of a largest anti-clique, computed using w, is at least equal to the inner product $\ell \cdot w$ of the weight vectors ℓ and w. (More generally, the max-max inequality, together with the assumption that $a^i \cdot b^j \leq 1$ for all rows a^i of A and b^j of B, characterizes antiblocking pairs of polyhedra.) In another and more recent paper [8] Lovász has proved the following simplification of Theorem 4.

Theorem 5. The graph G is perfect (pluperfect) if and only if (*) holds for all (0,1)-vectors $\ell = w$.

Thus a graph G is perfect (pluperfect) if and only if $\lambda(H)\omega(H) \geq |H|$ for all vertex-generated subgraphs H of G, where $|H|$ denotes the number of vertices of H.

In the context of anti-blocking pairs of matrices or polyhedra, Theorem 5 raises the following question. If we assume only that A is the (0,1)-incidence matrix of m pairwise noncomparable subsets of an n-set, that matrix B is a (0,1)-matrix whose rows satisfy $a^i \cdot b^j \leq 1$ for all $i = 1,\ldots,m$, $j = 1,\ldots,r$, and that

$$(**) \qquad (\max_i a^i \cdot w)(\max_j b^j \cdot w) \geq w \cdot w$$

for all (0,1)-vectors w, then it can be shown, using the Gilmore characterization of clique matrices [6], that A is the clique matrix of a graph G, B is the anti-clique matrix of G, and hence by Lovász's result, G is perfect, i.e., A and B are an anti-blocking pair of matrices. It would be interesting to know under what conditions on a (0,1)-matrix A and a nonnegative matrix B satisfying $a^i \cdot b^j \leq 1$ for all rows a^i of A and all rows b^j of B, the max-max inequality (*) for A and B can be replaced by (**), and still get the conclusion that A and B are an anti-blocking pair.

REFERENCES

1. C. Berge, Färbung von Graphen, deren sämtliche bzw. deren ungerade Kreise starr sind, Wiss. Z. Martin-Luther-Univ. Halle-Wittenberg Math. Natur. Reihe (1961), 114.

2. C. Berge, Graphes et Hypergraphes, Dunod, Paris (1970).

3. D. R. Fulkerson, Anti-blocking polyhedra, J. Comb. Th. 12 (1972), 50-71.

4. D. R. Fulkerson, Blocking and anti-blocking pairs of polyhedra, Math. Prog. 1 (1971), 168-194.

5. D. R. Fulkerson, The perfect graph conjecture and pluperfect graph theorem, Proceedings of the Second Chapel Hill Conference on Combinatorial Mathematics (1971).

6. P. Gilmore, Families of sets with faithful graph representation, Res. Note N. C. 184, IBM (1962).

7. L. Lovász, Normal hypergraphs and the perfect graph conjecture, to appear in Discrete Math.

8. L. Lovász, A characterization of perfect graphs, to appear in J. Comb. Th.

9. A. C. Tucker, The strong perfect graph conjecture for planar graphs, to appear.

Department of Operations Research
Cornell University
Ithaca, New York 14850

A Survey of Integer Programming Emphasizing Computation and Relations among Models

R. S. GARFINKEL AND G. L. NEMHAUSER

TABLE OF CONTENTS

1. Introduction

Integer programming deals with the class of mathematical programming problems in which some or all of the variables are required to be integers. We only consider the case in which both the objective function and constraints are linear so that the general model is

$$\text{(1)} \qquad \max z(x, v) = c_1x + c_2v, \qquad (x, v) \in S$$

where

$$S = \{(x, v) \mid A_1x + A_2v = b, \quad x \geq 0 \text{ integer}, \ v \geq 0\}$$

and $A_1(m\times n_1)$, $A_2(m\times n_2)$, $c_1(1\times n_1)$, $c_2(1\times n_2)$ and $b(m\times 1)$ are given. The problem (1) is called a mixed integer linear program (MILP). Two special cases of (1) are the integer linear program (ILP) in which $n_2 = 0$ and the linear program (LP) in which $n_1 = 0$. For ILP's, we assume that the matrices c_1, A_1 and b have integer elements. Rational elements can, of course, be transformed into integers.

If $S = \phi$ (empty set), (1) is said to be infeasible. If $S \neq \phi$, any $(x, v) \in S$ is called a feasible solution to (1). If there exists a least upper bound $\bar{z}$ such that

$$c_1x + c_2v \leq \bar{z} \quad \text{for all } (x, v) \in S$$

(1) is said to be <u>bounded.</u> In this case, there exists $(x^0, v^0) \in S$ such that

$$c_1 x^0 + c_2 v^0 = \bar{z}$$

and (x^0, v^0) is called an <u>optimal solution</u> to (1). If no such $\bar{z}$ exists, (1) is said to be <u>unbounded.</u>

Since the pioneering work of Ralph Gomory in the late 1950's the literature on integer programming has grown exponentially [more than 400 references are cited in Garfinkel and Nemhauser (1972b)]. We will not attempt a comprehensive survey. Instead, in an effort to complement existing surveys and books, we will:

1. describe some relationships among integer programming models and classify them according to computational complexity;
2. describe the most widely used algorithms;
3. report computational experience.

Although applications will not be discussed, they abound in government, industry and mathematics itself. To mention one problem in each of these areas, we cite political districting [Garfinkel and Nemhauser (1970)], facilities location [Spielberg (1969a), (1969b)] and graph coloring problems [Bessiere (1965)]. Woolsey (1972) discusses some of the difficulties encountered in applying integer programming in the real world.

Table 1 lists a number of other surveys and books, and comments on their emphases.

2. ILP Models and Some Relationships Among Them

The general ILP is written as

$$\text{(2)} \qquad \max cx,\ x \in S = \{x \mid Ax = b\ ,\ x \geq 0 \ \text{ integer}\}$$

where $A = (a_{ij})$ is $(m \times n)$. Some particular models are listed in Table 2.

Table 1

Author	Emphasis
Balinski (1965), article	A broad state-of-the-art survey.
Balinski and Spielberg (1969), article	A follow-up on Balinski (1965), it contains considerable material on enumerative methods and introduces the group-theoretic approach. A comprehensive list of about 200 references is included.
Hu (1969), book	This first book on integer programming emphasizes network models and cutting plane and group-theoretic methods for ILP's.
Greenberg (1971), book	A short book which has the novel feature of presenting some integer programming methods in a dynamic programming framework.
Geoffrion and Marsten (1972), article	A survey of general purpose methods that have been implemented with computational success. A general algorithmic framework for integer programming is given.
Garfinkel and Nemhauser (1972b), book	Comprehensive in all aspects (naturally this statement cannot be entirely objective). Numerous examples and exercises are given.

Table 2

Problem	Description
Binary ILP	x is a binary vector (each component of x is 0 or 1).
Set Covering, Partitioning and Packing [Garfinkel and Nemhauser (1972a)]	A binary ILP in which A is a binary matrix and b=e is a vector of ones. Given a set $I = \{1, \ldots, m\}$, column j of A represents the subset $\{i \mid a_{ij} = 1\} \subseteq I$. Covering: $Ax \geq e$ Partitioning: $Ax = e$ Packing: $Ax \leq e$.
Vertex Packing [Balinksi (1970)]	A set packing problem in which A is the edge-vertex incidence matrix of an undirected (loopless) graph. Thus each row of A has exactly two ones.
Edge Packing, also called edge matching [Edmonds (1965a)]	A set packing problem in which A is the vertex-edge incidence matrix of an undirected (loopless) graph. Thus each column of A has exactly two ones.
Single Commodity Flow [Ford and Fulkerson (1962)]	A is the vertex-edge incidence matrix of a (loopless) directed graph. Thus each column of A has precisely one (+1) and one (-1) .
Knapsack [Gilmore and Gomory (1966)]	$m = 1$ and $a_{1j} > 0$, $\forall$ j .
Binary Knapsack	A knapsack problem with x binary.
Integer Programming over Cones (ILPC), also called Asymptotic ILP [Gomory (1965)]	Delete the nonnegativity constraints in (2) on m variables whose corresponding columns of A are linearly independent.

We call an ILP <u>finite</u> if S is a finite set. For a finite ILP, there exists a vector u of suitably large integers such that

$$x \in S \Rightarrow x \leq u .$$

It is well-known that a finite ILP can be represented as a binary ILP using the transformation

$$(3) \qquad \begin{aligned} x_j &= \sum_{k=0}^{t_j} 2^k \delta_{kj}, \quad \forall j \\ \delta_{kj} &= 0,1 \quad \forall j,k \end{aligned}$$

where $2^{t_j} \leq u_j < 2^{t_j+1}$.

2.1 Transformation of Finite ILP's to Binary Knapsack Problems

Considerable attention has recently been devoted to the transformation of finite ILP's into binary knapsack problems [Bradley (1971), Elmaghraby and Wig (1970), Glover and Woolsey (1970) and Padberg (1970]. Without loss of generality, we consider a binary ILP with the two constraints

$$(4) \qquad \begin{aligned} a_1 x - b_1 &= 0 \\ a_2 x - b_2 &= 0 . \end{aligned}$$

By combining pairs of constraints, an m-constraint problem can be reduced to a binary knapsack problem. Assume that $a_1 \geq 0$, if not and $a_{1j} < 0$ let $x_j' = 1 - x_j$.
Define

$$\lambda = \sum_{j=1}^{n} a_{1j} - b_1$$

and consider the constraint

(5) $$(a_1 + \alpha a_2)x - b_1 - \alpha b_2 = 0$$

where α is any integer larger than λ. Clearly, any binary x which satisfies (4) also satisfies (5). If x^0 satisfies (5) then

(6) $$a_1 x^0 - b_1 = -\alpha(a_2 x^0 - b_2).$$

Now the right-hand side of (6) is an integer multiple of α, so the left-hand side must be as well. But from the choice of α, it follows that the left-hand side of (6) is smaller than α. Therefore

(7) $$a_1 x^0 - b_1 = 0$$

and (7) implies that x^0 satisfies (4).

Although the number of constraints is reduced by the transformation just given, the coefficients in (5) are generally large. Some control over the magnitude of the coefficients in (5) can be achieved by using a generalized version of the transformation in which both equations of (4) are multiplied by weights different from one.

2.2 Transformation of Finite ILP's to Packing Problems

Perhaps a more surprising result is that a finite ILP can be transformed into a set partitioning or set packing problem in which each column of A has no more than three ones, or into a vertex packing problem.† We begin with a binary ILP having $b \geq 0$ and first transform it into one in which A is a nonnegative matrix.

If there exists i such that $a_{ij} < 0$ replace $a_j x_j$ (a_j is the jth column of A) by

(8) $$a_j' x_{j1} + a_j'' x_{j2}$$

where

†Private communication (Jack Edmonds and Les Trotter, June 1972).

$$a'_{ij} = \max(0, a_{ij}) , \forall i$$

$$a''_{ij} = \min(0, a_{ij}), \forall i$$

and require $x_{j1} - x_{j2} = 0$ and $c_{j1} = c_{j2} = c_j/2$. Then using the transformation

$$y_{j2} = 1 - x_{j2} \tag{9}$$

we obtain a nonnegative constraint matrix.

The next transformation yields a binary constraint matrix with at most 3 ones per column. Replace a_j by the $(m \times \sum_{i=1}^{m} a_{ij})$ matrix $A'_j = (a'_{ik})$ where

$$a'_{ik} = \begin{cases} 0 \text{ if } 1 \leq k < t_{i-1,j} & \text{or } t_{ij} < k \leq t_{mj} \\ 1 \text{ if } t_{i-1,j} < k \leq t_{ij} \end{cases}$$

$$t_{ij} = \sum_{r=1}^{i} a_{rj}, \; i \geq 1 , \text{ and } t_{0j} = 0 . \tag{10}$$

Thus if

$$a_j = \begin{pmatrix} 2 \\ 0 \\ 3 \end{pmatrix} \text{ then } A'_j = \begin{pmatrix} 1 & 1 & 0 & 0 & 0 \\ 0 & 0 & 0 & 0 & 0 \\ 0 & 0 & 1 & 1 & 1 \end{pmatrix} .$$

Note that a unit vector requires no transformation.

Associate the variable x_{jk} with the kth column of A'_j and let

$$c_{jk} = \frac{c_j}{t_{mj}} , \forall k . \tag{11}$$

The new system is equivalent to the original if $x_j = x_{jk}$, $\forall k$. This condition is achieved with the constraints

$$x_{jk} + x'_{jk} = 1$$

(12) $$x_{j,k+1} + x'_{jk} = 1$$

$$x'_{jk} = 0, 1, \quad k = 1, \ldots, t_{mj} - 1 .$$

The x'_{jk} are assigned coefficients of zero in the objective function.

We can now assume that the system of equations has been reduced to

$$Ax = b$$

(13)

$$x \text{ binary}$$

where b is a nonnegative integer vector, A is a binary matrix and

$$\sum_{i=1}^{m} a_{ij} \leq 3 , \; \forall j .$$

It is convenient to view (13) as a problem on a graph with m vertices. The column a_j represents a loop, edge, or triangle (a cycle with three edges) depending on whether it has one, two, or three ones respectively. A feasible solution is then a collection of loops, edges and triangles such that the ith vertex has incidence b_i with the given collection. Note that A has been constructed so that each edge and triangle is incident to at most one vertex having $b_i \geq 2$. If a_ℓ is a loop (unit vector) such that $a_{i\ell} = 1$ and $b_i \geq 2$ add the constraint

$$x_\ell + x_s = 1$$

(14)

$$x_s = 0, 1$$

where the slack variable x_s is assigned a coefficient of zero in the objective function. The equation (14) yields a new vertex, so that a_ℓ represents an edge and the column corresponding to x_s represents a loop. Thus we can assume that every loop, edge and triangle is incident to at least one vertex having $b_i = 1$.

To achieve the condition $b_i = 1$, $\forall$ i, replace each vertex i such that $b_i \geq 2$, with b_i copies of itself. Then each edge and triangle that meet vertex i is replaced by b_i copies as shown in Figure 1.

The constraint system

$$\text{(15)} \qquad \begin{aligned} A'y &= e \\ y &\text{ binary} \end{aligned}$$

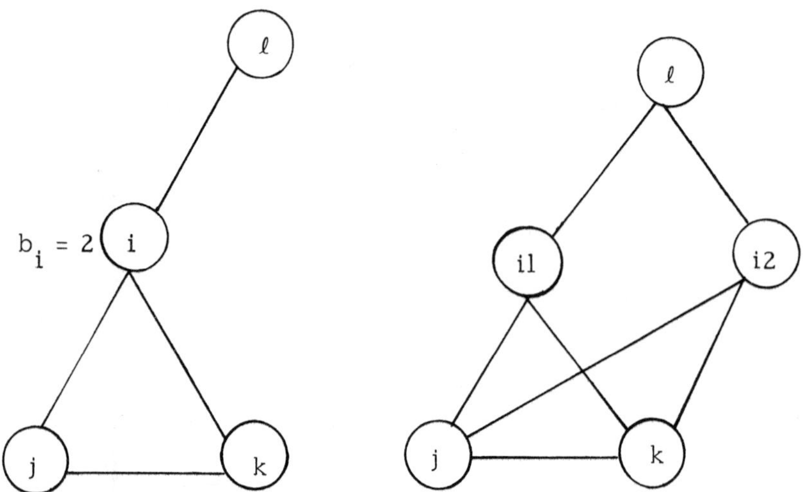

Figure 1

of a set partitioning problem with at most 3 ones per column has been obtained. If x_j is a variable in (13), incident with vertex t , where $b_t \geq 2$, its copies in (15) are denoted by $y_{j1}, \ldots, y_{jb_t}$ with $c_{jk} = c_j$, $\forall$ k. Since every loop, edge and triangle in (13) is incident with at least one vertex having $b_i = 1$, the condition

$$\sum_{k=1}^{b_t} y_{jk} \leq 1$$

is implied in (15). Thus given a solution to (15), a solution to (13) with the same objective value is obtained from

(16) $$x_j = \sum_{k=1}^{b_t} y_{jk} .$$

It is also clear that from a solution to (13), a solution to (15) that satisfies (16) can be obtained.

A set partitioning problem

(17) $$\begin{aligned} \max z(x) &= cx \\ Ax &= e \\ x &\text{ binary} \end{aligned}$$

can be further transformed into a set packing problem

(18) $$\begin{aligned} \max z'(x) &= c'x \\ Ax &\leq e \\ x &\text{ binary} . \end{aligned}$$

Let $\bar{z}$ be any upper bound on z for all binary x, $\underline{z}$ be a strict lower bound on z for all feasible solutions to (17) and $L = \bar{z} - \underline{z}$. Define

$$t_j = \sum_{i=1}^{m} a_{ij}$$

and

$$c'_j = c_j + Lt_j , \ \forall j .$$

Clearly, (18) has an optimal solution x^*. We will now show that:

(i) x^* feasible to (17) implies that x^* is optimal to (17) ;

(ii) x^* infeasible to (17) implies that (17) is infeasible.

Suppose (17) has the feasible solution $\hat{x}$ and let x' be a solution to (18) which is not feasible to (17). We have

(19) $$c'\hat{x} = c\hat{x} + Lm > \underline{z} + Lm$$

(20) $$c'x' \leq cx' + L(m-1) \leq \bar{z} + L(m-1) = \underline{z} + Lm .$$

From (19) and (20) we obtain

$$c'x' < c'\hat{x}$$

which implies that any optimal solution x^* to (18) is feasible to (17), provided that (17) has a feasible solution. Furthermore, for any feasible x to (17), we have

$$(c'-c)x = Lm = \text{constant}$$

so that x^* is an optimal solution to (17). The proof of (ii) is similar.

The problem (18) is transformed to a vertex packing problem by looking at the intersection graph defined by A [Edmonds (1962)]. The intersection graph contains a vertex for each column of A and vertices j and k are joined by an edge if there exists an i such that $a_{ij} = a_{ik} = 1$. Vertex j is assigned a value of c_j'. A subset of vertices is independent (contains no edges) and is therefore a solution to the vertex packing problem if and only if the corresponding columns of A do not intersect. The intersection condition is equivalent to $Ax \leq e$.

2.3 Single Commodity Flow Problems [Ford and Fulkerson (1962)]

Given a directed graph $G = (V, E)$, let a_i represent the amount of some commodity available at vertex i . (The availability is not necessarily positive). A flow $x = (x_{ij})$ is a set of numbers assigned to the edges that satisfy

(21) $$\sum_j x_{ij} - \sum_j x_{ji} \leq a_i , \quad \forall\ i \in V$$

(22) $$0 \leq x_{ij} \leq d_{ij} , \quad \forall\ (i,j) \in E$$

where the d_{ij}'s are nonnegative real numbers called capacities. A minimum cost flow is a flow that minimizes

(23) $$\sum_{(i,j)\in E} c_{ij}x_{ij} \; .$$

The problem of minimizing (23) subject to (21) and (22) is a highly structured LP called the minimum cost flow problem. If the a_i's and d_{ij}'s are integers, the ILP obtained by requiring x integer is essentially equivalent to the LP. The matrix A associated with the constraints (21) and (22) is totally unimodular (every square nonsingular submatrix has a determinant whose magnitude equals one) so that every basic feasible solution to the LP is integer.

A variety of well-known models including the shortest-path, transportation, assignment, and maximum flow problems are special cases of the minimum cost flow problem. To illustrate the shortest-path formulation, let c_{ij} be the length of edge (i,j) and define the length of a path to be the sum of the edge lengths over all edges in the path. Assume that the length of every cycle is nonnegative. The shortest-path problem is to find a path of minimum length between two specified vertices, say 1 and m. It is obtained from the minimum cost flow problem by setting $a_1 = 1$, $a_m = -1$, $a_i = 0$, otherwise and $d_{ij} = 1$, $\forall\,(i,j) \in E$. (Actually the capacity constraints can be ignored, since they will automatically be satisfied by some minimum cost flow.)

2.4 Formulation of Knapsack Problems as Shortest-Path Problems [Shapiro (1968c)]

Consider the knapsack problem

$$\max z(x) = \sum_{j=1}^{n} c_j x_j$$

(24) $$\sum_{j=1}^{n} a_j x_j = b$$

$$x \geq 0 \text{ integer.}$$

Without loss of generality assume $a_j \neq a_k$, $j \neq k$.

Let

$$E_j = \{(i,k) \mid k - i = a_j \,,\; 0 \leq i < k \leq b\}$$

and consider the directed graph $G = (V, E)$, where $V = \{0, 1, \ldots, b\}$ and

$$E = \bigcup_{j=1}^{n} E_j \,.$$

Let $-c_j$ be the length of each edge in E_j. The graph G is acyclic and has the structure shown in Figure 2. Any path in G from 0 to b yields a feasible solution to (24) by setting x_j equal to the number of edges in E_j in the path. Conversely, any feasible solution x to (24) can be interpreted as a path from 0 to b by choosing x_j edges from E_j, $j = 1, \ldots, n$. The length of such a path is equal to $-z$ for the corresponding solution to (24) so that a shortest path from vertex 0 to vertex b yields an optimal solution to the knapsack problem.

The binary knapsack problem cannot be treated using this formulation since there is no way to enforce $x_j \leq 1$. However, it can be represented as a shortest-path problem on a graph with nb vertices using a standard dynamic programming formulation.

2.5 Formulation of the ILPC as a Shortest-Path Problem [Gomory (1965), Shapiro (1968a), Hu (1970)]

Consider the LP

$$\begin{aligned} \max\ z(x) &= cx \\ Ax &= b \\ x &\geq 0 \end{aligned} \tag{25}$$

and suppose that A has full row rank. Permute the columns of A so that $A = (B, N)$, where B is nonsingular, and rewrite (25) as

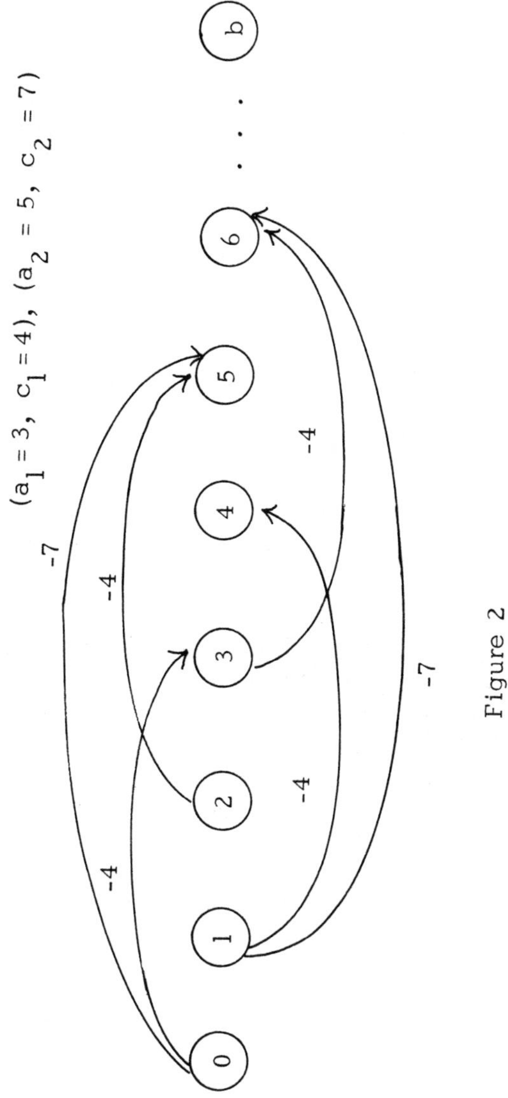

Figure 2

$$\max z(x) = c_B x_B + c_N x_N$$

(26)
$$Bx_B + Nx_N = b$$

$$x_B \geq 0\ ,\quad x_N \geq 0\ .$$

In (26), the matrix B is called a basis matrix, x_B is called the vector of basic variables and x_N is called the vector of nonbasic variables. Solving (26) for z and x_B in terms of x_N yields the representation

$$\max z(x) = c_B B^{-1} b - (c_B B^{-1} N - c_N) x_N$$

(27)
$$x_B = B^{-1}b - B^{-1}Nx_N$$

$$x_B \geq 0\ ,\quad x_N \geq 0\ .$$

The solution

(28)
$$z = c_B B^{-1} b,\ x_B = B^{-1}b,\ x_N = 0$$

is called a basic solution to the LP. If, in addition, $B^{-1}b \geq 0$, the basic solution is called primal feasible; and, if $c_B B^{-1} N - c_N \geq 0$, the basic solution is called dual feasible. An essential result of linear programming is that a basic solution which is primal and dual feasible is optimal.

Now consider a dual feasible basic solution and augment (26) by the condition x integer. This yields a representation of the ILP as

$$\max z(x) = c_B x_B + c_N x_N$$

$$Bx_B + Nx_N = b$$

(29)
$$x_B \geq 0 \quad \text{integer}$$

$$x_N \geq 0 \quad \text{integer}\ .$$

Deleting $x_B \geq 0$ from (29), we obtain the ILPC

$$\min z'(x_N) = (c_B B^{-1} N - c_N) x_N$$
(30) $$B^{-1}b - B^{-1}Nx_N \quad \text{integer}$$
$$x_N \geq 0 \quad \text{integer}.$$

In (30) the variables x_B have been eliminated using the equation in (27), and the constant $c_B B^{-1} b$ has been dropped. The practical importance of the ILPC model is that if x_N^* is an optimal solution to (30) and if

(31) $$x_B^* = B^{-1}b - B^{-1}Nx_N^* \geq 0$$

then (x_B^*, x_N^*) is an optimal solution to the ILP (29). By choosing B so that the solution (28) is primal and dual feasible, frequently some optimal solution to (30) will satisfy (31). In fact, it can be shown that if the solution $B^{-1}b$ is not degenerate then, for large enough b, this condition always holds.

To transform (30) into a shortest-path problem we apply the result of Smith [see MacDuffee (1940)] that for non-singular, integer B there exist unimodular matrices R and C such that

(32) $$RBC = \Delta$$

where Δ is a diagonal matrix with diagonal elements $(\delta_1, \ldots, \delta_m)$ satisfying:

1. $\delta_i > 0, \quad i = 1, \ldots, m.$

2. δ_i is a divisor of $\delta_{i+1}, \quad i = 1, \ldots, m-1.$

3. $\prod_{i=1}^{m} \delta_i = |\text{determinant } B| = D.$

Multiplying

(33) $$Bx_B + Nx_N = b$$

by R and then substituting

$$RB = \Delta C^{-1}$$

yields

$$\Delta C^{-1}x_B + RNx_N = Rb . \tag{34}$$

Since C^{-1} is integer and unimodular and x_B is unconstrained, we can substitute

$$y = C^{-1}x_B$$

into (34) to obtain

$$\Delta y + RNx_N = Rb \tag{35}$$

and the result that (y^*, x_N^*) is an integer solution to (35) if and only if (Cy^*, x_N^*) is an integer solution to (33). The ith row of (35) is

$$\delta_i y_i + R_i Nx_N = R_i b \tag{36}$$

where R_i is the ith row of R. Equation (36) is equivalent to

$$R_i Nx_N \equiv R_i b \pmod{\delta_i} \tag{37}$$

which means that $R_i Nx_N - R_i b$ is an integer multiple of δ_i. Thus the ILPC can be written

$$\min z'(x_N) = (c_B B^{-1}N - c_N)x_N$$

$$R_i Nx_N \equiv R_i b \pmod{\delta_i}, \quad i = 1, \ldots, m \tag{38}$$

$$x_N \geq 0 \quad \text{integer.}$$

Let $N = (a_1, \ldots, a_r)$ and consider a directed graph $G = (V, E)$ with $V = (1, \ldots, D)$. The D vertices are placed on an m-dimensional integer grid so that the coordinates of an arbitrary vertex, say k , are

$$(p_1^k, \dots, p_m^k), \quad 0 \leq p_i^k \leq \delta_i - 1, \quad \forall\, i .$$

Let

$$E_j = \{(k,t) \mid R_i a_j + p_i^k \equiv p_i^t \pmod{\delta_i}, \quad i = 1, \dots, m\}$$

and

$$E = \bigcup_{j=1}^{r} E_j .$$

Each edge in E_j is assigned the length $c_B B^{-1} a_j - c_j$.

All feasible solutions to (38) can be represented as paths from the vertex at $(0, \dots, 0)$ to the vertex at $(R_1 b \pmod{\delta_1}, \dots, R_m b \pmod{\delta_m})$ and conversely, by setting x_j equal to the number of edges in E_j in the path. The length of such a path is $(c_B B^{-1} N - c_N) x_N$ so that a shortest path yields an optimal solution to the ILPC.

2.6 Transforming an MILP to an ILP [Benders (1962)]

Substituting any nonnegative integer vector x' into the MILP (1) yields the LP

$$\begin{aligned} \max\ z(x', v) &= c_1 x' + c_2 v \\ A_2 v &\leq b - A_1 x' \\ v &\geq 0 . \end{aligned} \tag{39}$$

The dual of (39) is

$$\begin{aligned} \min\ w(x', u) &= c_1 x' + u(b - A_1 x') \\ u A_2 &\geq c_2 \\ u &\geq 0 . \end{aligned} \tag{40}$$

The constraint set

$$P = \{u \mid uA_2 \geq c_2 \ , \ u \geq 0\}$$

is independent of x'. Let

$$T = \{u^t \mid u^t \text{ an extreme point of } P\}$$

and

$$Q = \{y^q \mid y^q \text{ an extreme ray of } P\}.$$

Note that if T is empty, there is no optimal solution to (1). By expressing all feasible solutions to (40) in terms of the elements of T and Q, it is possible to reformulate (1) as

$$\max u_0$$

$$\begin{aligned} u_0 &\leq c_1 x + u^t(b - A_1 x) \ , \quad u^t \in T \\ 0 &\leq \qquad\quad y^q(b - A_1 x), \quad y^q \in Q \end{aligned} \tag{41}$$

$$x \geq 0 \text{ integer.}$$

The problem (41) is an MILP having one continuous variable u_0. It has been customary to refer to (41) as an ILP, although any method for solving ILP's would have to be modified slightly to solve it. Since there is a constraint in (41) for each extreme point and extreme ray of P, (41) generally has an enormous number of constraints. However, as shown by Benders, it can be solved iteratively by generating the constraints only if and when they are needed.

2.7 Transforming an ILP into an LP

Any finite ILP can be written, at least in principle, as an LP by augmenting $S' = \{x \mid Ax = b, \ x \geq 0\}$ by those linear constraints that define the convex hull of feasible solutions to (2). If A in (2) is totally unimodular then no augmentation is necessary, since it is clear that every extreme point of S' is integer for all integer vectors b. When

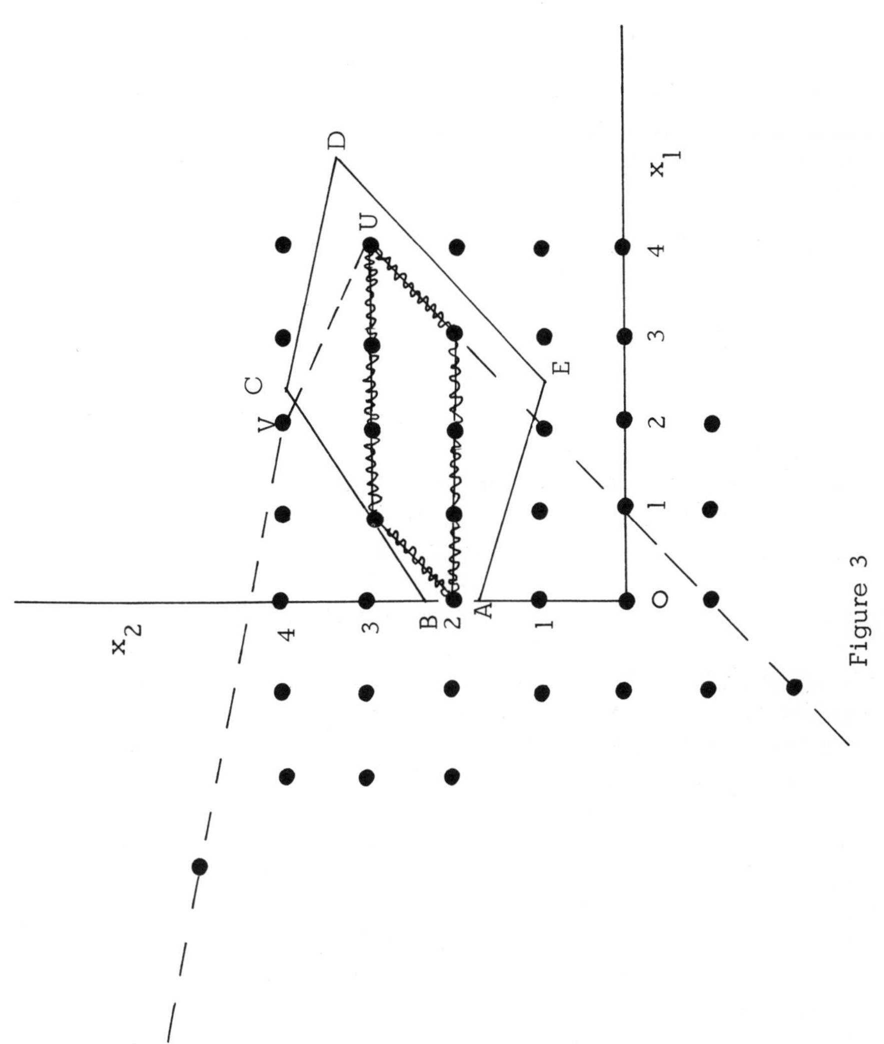

Figure 3

$S' = \{x \mid Ax \leq b\ ,\ x \geq 0\}$ the converse is also true [Hoffman and Kruskal (1958), Veinott and Dantzig (1968)]. That is, if all extreme points of S are integer for all integer vectors b, then A is totally unimodular.

Only in very few cases have explicit representations of the convex hull of a set of integer points been obtained. One of the most notable of these is edge packing [Edmonds (1965b)], where it has been shown that the convex hull of integer solutions is given by

$$Ax \leq e$$

$$\sum_{j \in E(C)} x_j \leq \frac{|C| - 1}{2}, \ \forall\ C \subseteq V, \text{ such that } |C| \geq 3 \text{ and odd}$$

$$x \geq 0$$

where A is the vertex-edge incidence matrix of G, E(C) is the subset of edges that are incident to two vertices of C and $|C|$ denotes the cardinality of C .

Gomory (1967), (1969) has characterized the extreme points and faces of the polyhedra for ILPC's. This characterization is based on the result that the columns of an ILPC correspond to elements of a finite abelian group which is the direct sum of cyclic groups of order δ_i, $i = 1, \ldots, m$, where the δ_i are the diagonal elements of Δ in (32). The faces of these polyhedra, called <u>corner</u> polyhedra, can be obtained from the basic feasible solutions to an LP. In Figure 3, a two-dimensional case is illustrated. Feasible solutions to the constraints $Ax \leq b\ ,\ x \geq 0$ are points contained within the polyhedron ABCDE. The convex hull of solutions to $Ax \leq b\ ,\ x \geq 0$ integer is given by the polyhedron bounded by wiggly lines. The corner polyhedron determined from the extreme point D is the unbounded region within the dashed lines and has extreme points U and V . Clearly, the corner polyhedron contains the convex hull of feasible solutions to the ILP.

3. Computational Complexity of Integer Programming Problems

The experimental approach is the obvious way to evaluate the computational difficulty of integer programming problems. By running a number of "test" problems on existing algorithms, we can empirically compare the algorithms and collect information on computation time as a function of structure and parameters such as number of variables and constraints. We will report results of this kind in Section 7.

In this section we briefly describe some recent results on a theoretical way of measuring computational complexity. By complexity we mean the number of basic computations required, measured by the number of additions and comparisons.

Given a class of problems K that are solvable by a finite algorithm, and the class H of all possible finite algorithms for finding optimal solutions to problems in K, let $t(x, y)$ be the number of computations required to solve problem x by algorithm y. Since t is generally an unknown function, for a given algorithm y, we seek an upper bound of the form

$$\bar{z}(y) \geq \sup_{x \in K} t(x, y)$$

and lower bounds of the form

$$\underline{z} \leq \inf_{y \in H} \sup_{x \in K} t(x, y) .$$

If $\bar{z}(y^*) = \underline{z}$, the algorithm y^* is called optimal for K.

Even if $\bar{z}(y)$ is a tight upper bound, it can represent a very pessimistic measure of performance of algorithm y, since it may reflect results for pathological rather than typical problems in K. For example, the number of iterations required by the simplex method to solve most LP's that arise in practice has been empirically established to be a linear function of the number of constraints [Wolfe and Cutler (1963)]. However, it has recently been shown [Klee and Minty (1970)] that the number of iterations can grow exponentially.

Not much is known about sharp lower bounds for integer programming problems except in some obvious cases. In problems defined on n-vertex graphs, if every edge is a candidate for membership in an optimal solution, then each edge will have to be examined at least once. We can conclude that a lower bound on complexity is given by a polynomial in n of degree 2 or $0(n^2)$.

Dijkstra's algorithm (1959) for finding shortest paths between a specified vertex and all others on graphs with positive edge lengths is $0(n^2)$ and therefore is optimal with respect to the order of complexity. There are also $0(n^2)$ algorithms for finding shortest paths on acyclic graphs. Various algorithms for the more general case of the shortest-path problem are $0(n^3)$, but are not necessarily optimal.

Since proving optimality requires the knowledge of a tight lower bound, very little is known about the optimality of integer programming algorithms. For this reason, we consider another criterion for measuring the quality of an algorithm.

Following Edmonds (1965a), an algorithm will be called <u>good</u> for some class of problems, if an upper bound on its complexity is a polynomial in the length of the encoding of the problem data. This idea has been embodied in a theory of computational complexity that allows one to classify problems according to a precise measure of difficulty.

A standard encoding of problem data is a finite string of zeroes and ones. The length of the string is

$$L = \text{number of zeroes} + \text{number of ones}.$$

A problem is said to be in the class P (polynomial) with respect to its encoding if there is an algorithm whose number of computations is polynomial in L.

A proper encoding is essential for obtaining meaningful results. For example, consider a binary ILP and suppose the encoding included the data c, b and A together with the 2^n binary vectors. Also, consider an algorithm that simply evaluates the feasibility and objective value of each vector in turn and saves the best feasible solution. With respect to this encoding the algorithm is good, but one could hardly call it an efficient algorithm for solving binary ILP's. There is, of course, a much more natural and smaller encoding for a binary ILP. Therefore, solution by this naive algorithm is not a sufficient condition for membership in P .

Graphs can be encoded by their adjacency or incidence matrix, integers by their binary representation and vectors and matrices as lists of integers. It is important to distinguish between the binary encoding of the integer m for which $L = \log_2 m$, and the unary encoding (only one symbol) for which $L = m$. To appreciate the distinction, consider the knapsack problem which can be represented as a shortest-path problem on a directed, acyclic graph with b+1 vertices. If the knapsack problem is encoded as a shortest-path problem, one can conclude that there is a good algorithm of complexity $0((b+1)^2)$. However if the 2n+1 pieces of integer data are encoded in their binary representation, L grows logarithmically with b and an $0(b^2)$ algorithm is not polynomial in L . Therefore, the shortest-path formulation does not qualify the knapsack problem for membership in P .

Besides the shortest-path problem, the maximum-flow and assignment problems there are other minimum-cost flow problems that belong to P . Suppose we are given a directed graph with integer capacities on the edges and wish to maximize the flow between two specified vertices called the source (s) and the sink (t). Ignoring the orientation of the edges, consider any path between s and t . On such a path an edge is called <u>forward</u> if it obeys its orientation and <u>reverse</u> if it violates its orientation. Given an arbitrary integer feasible flow x , a flow augmenting path is a path such that

1. $x_{ij} < d_{ij}$ (the capicity) for all forward edges in the path
2. $x_{ij} \geq 1$ for all reverse edges in the path.

The total flow from s to t can be increased if and only if a flow augmenting paths exists. Ford and Fulkerson (1962) have given a good algorithm that either finds a flowing path or establishes that none exists. This is not sufficient, however, for a good maximum flow algorithm

Consider the example in Figure 4, where M is a large integer. If any flow augmenting path could be selected, the algorithm could take 2M steps using the sequence of paths (1, 2, 3, 4), (1, 3, 2, 4), (1, 2, 3, 4), (1, 3, 2, 4).... However, the maximum flow problem is in P, since Edmonds and Karp (1972) have recently found an $0(n^3)$ algorithm, independent of the capacity data. At each iteration, this algorithm finds a shortest augmenting path, where length is measured by the number of edges a path contains.

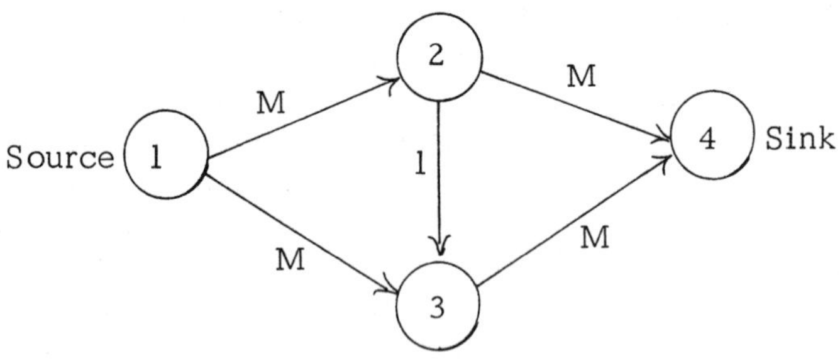

Figure 4

Given m_1 men and m_2 jobs $(m_1 + m_2 = n)$, and a value c_{ij} of man i doing job j, the assignment problem is to find an allocation of men to jobs such that each man does at most one job, each job is done by at most one man and total value is maximized. This problem can be interpreted as a minimum cost flow problem or as an edge packing problem on a bipartite graph. There are algorithms of complexity $0(n^3)$ for solving it. An $0(n^3)$ algorithm of Edmonds (1965a) solves the edge packing problem on an arbitrary n-vertex graph. Lawler (1971) indicates the $0(n^3)$ bound rather than $0(n^4)$ as had generally been supposed.

Some recent results of Cook (1971) and Karp (1972) cast doubt on the existence of good algorithms for general ILP's. Since a rigorous presentation of their results requires a background in the theory of computation, we only sketch the basic ideas.

Suppose we were to solve a binary ILP by enumerating the 2^n binary n-vectors on a tree (approaches of this type will be discussed in the next section). If we had a computer which could duplicate itself whenever, at any stage of the enumeration, it was necessary to consider both alternatives ($x_j = 0$ or 1), and the complexity at each stage was a polynomial in n, then, since the depth of tree is n, the complexity for complete enumeration would be a polynomial in n. The class of problems that can be solved by a good algorithm on such a computer is called NP (nondeterministic polynomial). Problems are in this class if and only if they can be solved by an enumeration algorithm on a tree whose depth is polynomial in L. Of course, $P \subseteq NP$. The class NP is quite extensive and includes all finite ILP's and LP's, and numerous well-known combinatorial optimization problems including the traveling salesman problem.

A member of NP is the satisfiability problem denoted SAT. Suppose we are given n statements $(P_1, \ldots, P_n)$ each of which is either true or false and a compound statement which is the conjunction of m clauses, where each clause is a disjunction of a subset of the statements and their negations. SAT is to determine if there exist truth values for the individual statements such that the compound statement is true. Such truth values exist if and only if there is a binary vector $x = (x_1, \ldots, x_n)$ such that

$$\sum_{j \in Q_i} x_j + \sum_{j \in \bar{Q}_i} (1 - x_j) \geq 1, \quad i = 1, \ldots, m \tag{42}$$

has a solution, where $Q_i \cup \bar{Q}_i$ is the index set of statements that appear in clause i and $\bar{Q}_i$ is the index set of negated statements in clause i. Then $x_j = 1$ means that P_j is true and $x_j = 0$ means that P_j is false.

Cook's main result is that every problem in NP is <u>reducible</u> to SAT. Reducibility has a precise definition which, unfortunately, is beyond the scope of our introduction. Loosely, however, problem L is reducible to problem M if there exists a function f computable in a polynomial number of steps, such that x is an encoding for L if and only if $f(x)$ is an encoding for M.

A corollary of Cook's theorem is

$$P = NP \iff SAT \in P .$$

Thus if there were a good algorithm for SAT, there would be a good algorithm for every problem that can be solved by enumeration on a tree whose depth is polynomially bounded. One does not have to be very pessimistic to believe that SAT $\notin$ P. However another reasonable point of view is that SAT $\in$ P, but a good algorithm requires a polynomial of high degree. It appears that very little effort has gone into constructing good algorithms of complexity $0(n^M)$, where M is very large. It is not even clear that the human mind is capable of doing so.

Karp has extended Cook's results by showing that numerous problems in NP are equivalent to SAT in the sense that SAT can be reduced to them. These problems are called complete. Karp's list of complete problems include binary ILP's (by virtue of (42)), the binary knapsack problem, set covering and partitioning and vertex packing. It is not known if LP's are complete. If any of the complete problems are in P, it follows that P = NP.

4. Enumeration Algorithms

4.1 General Algorithm

In this section we present a general enumeration algorithm [Bertier and Roy (1964), Agin (1966), Lawler and Wood (1966), Mitten (1970)] for the mathematical programming problem

$$\max z(x) , \quad x \in S_0 . \tag{43}$$

Solutions to (43) are enumerated by means of a tree, with vertex 0 corresponding to (43), and vertex j corresponding to

$$\max z(x) , \quad x \in S_j \subset S_0 . \tag{44}$$

Separation

The tree is generated by separating S_j into a set S_j^* of subsets of S_j where

$$\bigcup_{T \in S_j^*} T = S_j \ .$$

The set S_j^* is called a separation of S_j. In many realizations the separation has the desirable property of being a partition. Every vertex of the enumeration tree which emanates from vertex j is called a successor of vertex j, and corresponds to one of the elements of S_j^*.

Bounding

Let

$$z_j^* = \begin{cases} z(x^*(j)) & \text{if } x^*(j) \text{ is an optimal solution to (44)} \\ -\infty & \text{if } S_j = \emptyset \\ \infty & \text{if (44) is unbounded.} \end{cases}$$

An upper bound $\bar{z}_j \geq z_j^*$ may be calculated by considering the relaxation of (44)

$$\max z(x) \ , \quad x \in T_j \supseteq S_j \ . \tag{45}$$

Let

$$\bar{z}_j = \begin{cases} z(x^0(j)) & \text{if } x^0(j) \text{ is an optimal solution to (45)} \\ -\infty & \text{if } T_j = \emptyset \\ \infty & \text{if (45) is unbounded.} \end{cases}$$

Much of the art of branch and bound involves choosing T_j effectively. It is generally necessary to compromise between the sharpness of the bound and the difficulty in solving (45).

A lower bound $\underline{z}_j \leq z_j^*$ is available if any $x' \in S_j$ is known, and is given by $z(x')$. Since $S_j \subseteq S_0$, it also follows that $\underline{z}_0 = z(x')$ is valid.

Fathoming

A vertex j of the enumeration tree is said to be fathomed if it is known that S_j does not contain any point with objective larger than $\underline{z}_0$. Thus if $\underline{z}_0$ is the value of the best known solution in S_0, then vertex j is fathomed if $\bar{z}_j \leq \underline{z}_0$. Clearly, no further enumeration is required from a fathomed vertex. Any vertex which is not fathomed is said to be live.

Algorithm

Step 1: (Initialization). Begin at the live vertex 0, and let $\bar{z}_0$ and $\underline{z}_0$ be upper and lower bounds on z (possibly $+\infty$ and $-\infty$ respectively). Go to Step 2.

Step 2: (Branching). If no live vertices exist, go to Step 7. Otherwise, choose a live vertex j. If (45) has been solved at vertex j, go to Step 3; otherwise go to Step 4.

Step 3: (Separation). Choose a separation S_j^*, which determines the successors of vertex j. Go to Step 2.

Step 4: (Bounding). Solve (45). If $T_j = \phi$, vertex j is fathomed and we go to Step 2. If (45) has an optimal solution $x^0(j)$, let $\bar{z}_j = z(x^0(j))$ and go to Step 5. If (45) is unbounded, let $\bar{z}_j = \infty$ and go to Step 2.

Step 5: If $x^0(j) \in S_j$, let $z_0 = \max\{\underline{z}_0, z(x^0(j))\}$. Go to Step 6.

Step 6: (Fathoming). Any vertex i with $\bar{z}_i \leq \underline{z}_0$ is fathomed. Go to Step 2.

Step 7: (Termination). If $\underline{z}_0 > -\infty$, that solution which yielded $\underline{z}_0$ is optimal. If $\underline{z}_0 = -\infty$, either $S_j = \phi$ or (43) is unbounded.

Finiteness

Sufficient conditions for finiteness of the tree are that $|S_j^*|$ be finite for all j and that there exists a function

$$f: \text{subsets of } S \rightarrow \text{nonnegative integers}$$

having the property that if $f(S_j) > 1$

$$(46) \qquad S_k \in S_j^* \Rightarrow f(S_k) < f(S_j) .$$

If, in addition, vertex j can be fathomed whenever $f(S_j) \leq 1$, the tree must be finite. For some specializations of the general algorithm we will show that such functions exist.

ε-Optimality

One important feature of branch and bound algorithms is that they produce both feasible solutions and bounds on optimal solutions. Thus, the user always has the option of regarding vertex j as fathomed if an upper bound $\bar{z}_j$ is produced such that $\bar{z}_j \leq \underline{z}_0 + \varepsilon \underline{z}_0$, where $\varepsilon \geq 0$ is determined by the user. The use of this option can significantly cut down the tree size.

4.2 Specialization of the General Algorithm to MILP's via Linear Programming

[Land and Doig (1960), Dakin (1965), Roy et al. (1970), Tomlin (1970), (1971)]

A specialization of the general branch and bound algorithm to the MILP (1) has subproblems (44) with the constraint set

$$(47) \quad S_j = \{(x, v) \,|\, A_1 x + A_2 v = b, \ \alpha^j \leq x \leq \beta^j, \ x \text{ integer}, \ v \geq 0\}$$

where α^j and β^j are nonnegative integer vectors. In the relaxation (45), T_j is given by

$$(48) \qquad T_j = \{(x, v) \,|\, A_1 x + A_2 v = b, \ \alpha^j \leq x \leq \beta^j, \ v \geq 0\}$$

so that (45) is an LP called the <u>corresponding LP</u> to (44). When the LP (45) is solved at vertex j , separation is required only if the optimal solution $(x^0(j), v^0(j))$ to (45) is not in S_j . In that case, some basic variable x_i has value $y_i = [y_i] + f_i$, where $f_i > 0$ is the fractional part of y_i .

Thus an appropriate separation is

(49) $$S_j^* = \{S_j \cap \{(x,v) | x_i \leq [y_i]\}, \; S_j \cap \{(x,v) | x_i \geq [y_i] + 1\}\}.$$

The separation (49), generates the vectors α^j and β^j in (47) and (48). Initially, unless tighter bounds are known, one can take $\alpha^0 = (0,\ldots,0)$ and $\beta^0 = (\infty,\ldots,\infty)$. Since the constraints imposed by (49) are simply bounds on the variables, the LP with constraint set (48) can be solved by the dual simplex algorithm for bounded variables. In fact, it may not be necessary to actually solve the LP. If the objective function reaches $\underline{z}_0$ during the course of dual simplex calculations, the vertex can be fathomed.

Finiteness of the algorithm can be guaranteed if β^0 is a vector of integers. In that case the function f of (46) can be taken to be

(50) $$f(S_k) = \prod_{j=1}^{n} (\beta_j^k - \alpha_j^k + 1)\,.$$

Clearly, f as given by (50) satisfies (46). Also, $f(S_k) = 1$ implies $x(k) = \alpha^k = \beta^k$, so that vertex k can be fathomed by inspection.

Branching

The separation (49) leaves open the question of which basic variable to branch on, if more than one has $f_i > 0$. In operating branch and bound codes, this decision is generally made based on <u>penalties</u> which are calculated for all possible separations. These penalties are lower bounds on the decreases in the objective function of (45) and can be calculated in a number of ways. One rule that has proven effective is to calculate, for every i such that $f_i > 0$, an "up penalty" U_i and a "down penalty" D_i, which are the decreases in objective function at the next dual simplex iteration produced if the constraint $x_i \geq [y_i] + 1$ or $x_i \leq [y_i]$ respectively, is appended. Then, if

$$P_k = \max_{i, f_i > 0} \max\{D_i, U_i\}$$

we separate on x_k and branch up if $P_k = D_k$ and down if $P_k = U_k$.

It also seems most common in branch and bound algorithms to branch to one of the successor vertices of the vertex under consideration. Thus, one has the advantage of working with a problem currently in the computer core. If the current vertex is fathomed one "backtracks" along the unique path to vertex zero until the first vertex having a live successor is encountered.

However, some authors use additional criteria for branching. For instance in Roy et al. (1970) this decision is made based on information about the live vertices, such as upper bounds, position in the enumeration tree, and number of variables currently integer. Clearly if the objective were to minimize the tree size, the branching criterion would be to choose the live vertex with the greatest upper bound.

4.3 Implicit Enumeration--Another Realization of Branch and Bound

[Balas (1965), Glover (1965), Geoffrion (1967), Lemke and Spielberg (1967)]

Implicit enumeration is the name for a branch and bound algorithm for the binary ILP

(51) $$\max z(x) = cx \ , \ x \in S_0$$

where

$$S_0 = \{x \,|\, Ax \leq b, \quad x \text{ binary}\} \ .$$

Without loss of generality we can take $c \leq 0$ in (51), since the transformation $x_j' = 1 - x_j$ achieves this for $c_j > 0$. At vertex k of the enumeration tree, the index set $J = \{1, \ldots, n\}$ of the variables is partitioned into

$$J = S_k^+ \cup S_k^- \cup F_k$$

where

$$S_k^+ = \{j \mid x_j = 1\}$$

$$S_k^- = \{j \mid x_j = 0\}$$

$$F_k = \{j \mid x_j \text{ is free to be zero or one}\} .$$

Then the problem at vertex k is

$$\max z_k(x) = \sum_{j \in F_k} c_j x_j + \hat{z}_k$$

(52) $$\sum_{j \in F_k} a_{ij} x_j \leq s_i^k , \quad i = 1, \ldots, m$$

$$x_j = 0, 1, \quad j \in F_k$$

where

$$\hat{z}_k = \sum_{j \in S_k^+} c_j$$

and

$$s_i^k = b_i - \sum_{j \in S_k^+} a_{ij} .$$

Consider the relaxation of (52)

(53) $$\max z_k(x), \ x \in T_k = \{j \mid x_j = 0, 1, \ j \in F_k\} .$$

Since $c \leq 0$ it follows that an optimal solution to (53) with value $\hat{z}_k$ is $x_j = 0$, $j \in F_k$. Thus, if $s^k = (s_1^k, \ldots, s_m^k) \geq 0$, or if $\hat{z}_k \leq \underline{z}_0$, vertex k is fathomed.

If vertex k is not fathomed, the separation is

$$S_k^* = \{S_k \cap \{x \mid x_j = 1\}, \ S_k \cap \{x \mid x_j = 0\}\} \qquad \text{for some } j \in R_k$$

where

$R_k = \{j \mid a_{ij} < 0$ for at least one i such that $s_i^k < 0, j \in F_k\}$.

A common rule is to choose $t \in R_k$ where t minimizes the resulting infeasibility given by

$$\sum_{i=1}^{m} \max\{0, -s_i^k + a_{ij}\} .$$

One then branches to the successor of vertex k corresponding to $x_t = 1$.

Fathoming Tests and Surrogate Constraints

Vertex k may also be fathomed if

$$\sum_{j \in F_k} \min\{0, a_{ij}\} > s_i^k \tag{54}$$

for any i. Other tests similar to (54), and based on the binary requirement on the variables have been exploited [Glover (1965), Glover and Zionts (1965), Fleischmann (1967), Petersen (1967)]. It has also proved profitable to perform these tests first on a <u>surrogate constraint</u> of the form

$$\sum_{i=1}^{m} \sum_{j \in F_k} u_i a_{ij} x_j \leq \sum_{i=1}^{m} u_i s_i^k \tag{55}$$

where $u_i \geq 0$, $i = 1, \ldots, m$ [Glover (1965), (1968b), Balas (1967), Geoffrion (1969)]. Clearly, if (55) has no binary solution, neither does (52).

It is important to choose a vector $u = (u_1, \ldots, u_m)$ so that the potential for fathoming is great. Let $S_k(u)$ be the set of binary solutions to (55) and

$$g(u) = \max z_k(x) , \quad x \in S_k(u) . \tag{56}$$

Then a reasonable definition of the strongest surrogate constraint is one generated by u^*, where u^* minimizes $g(u)$ over all nonnegative u. Unfortunately, this definition does not lend itself to a straightforward technique for computing the strongest surrogate. However, an estimate of u^* which is readily computed can be derived by relaxing the integrality requirement on x in (56), to $0 \leq x_j \leq 1$, $j \in F_k$. In this case it can be shown that the strongest surrogate constraint is given by u^* where (u^*, w^*) solves the LP

$$\min \sum_{i=1}^{m} s_i^k u_i + \sum_{j \in F_k} w_j$$

$$\sum_{i=1}^{m} a_{ij} u_i + w_j \geq c_j , \quad j \in F_k$$

$$u_i, w_j \geq 0 \quad \forall\, i, j$$

which is dual to the LP obtained from (52) by relaxing the integrality requirements on x. Thus, to calculate the strongest surrogate constraint at vertex k, one solves the LP corresponding to (52). In so doing, all of the power of the bounds derived in the previous LP-based approach is available.

4.4 Branch and Bound and the ILPC [Shapiro (1968b), Gorry and Shapiro (1971)]

For the ILP, a branch and bound algorithm has also been derived using the relaxation (30) with an optimal LP basis B. The relaxation at vertex 0 of the enumeration tree is

$$\max z(x_N), \quad x_N \in T_0$$

where

$$z(x_N) = c_B B^{-1} b - (c_B B^{-1} N - c_N) x_N$$

and

$$T_0 = \{x_N \mid B^{-1} N x_N \equiv B^{-1} b \pmod 1),\ x_N \geq 0 \text{ integer}\} .$$

Note that T_0 is the constraint set of the ILPC (30).

Letting $x_N = (x_1, \ldots, x_r)$, at vertex k the relaxed problem is

$$(57) \qquad \max z(x_N) \, , \; x_N \in T_k = T_0 \cap \{x_N \mid x_N \geq p^k\}$$

where $p^k = (p_1^k, \ldots, p_r^k)$ is a vector of nonnegative integers. The problem (57) can be modeled as a shortest-path problem in the same manner as (30) by making the change of variables $x_N' = x_N - p^k$.

Separation

The vector p^k in (57) is determined by the separation

$$(58) \qquad S_k = \bigcup_{j=1}^{r} S_{k(j)}$$

where

$$S_{k(0)} = S_k \cap \{x_N \mid x_N = p^k\}$$

and

$$S_{k(j)} = S_k \cap \{x_N \mid x_j \geq p_j^k + 1\} \, , \quad j = 1, \ldots, r \, .$$

Although the separation (58) is not a partition, it can be shown that it is not necessary to consider the problems at vertices $k(j)$, $j = 0, \ldots, j(k) - 1$, where

$$j(k) = \begin{cases} \max\{j \mid j \in R_k\} & \text{if } R_k \neq \emptyset \\ 1 & \text{if } R_k = \emptyset \end{cases}$$

and $R_k = \{j \mid p_j^k > 0\}$. For $j = 0$, either $S_{k(0)} = \emptyset$, or $S_k(0) = \{p^k\}$. In the former case, there is no point in branching to vertex $k(0)$, and in the latter case, by virtue of the monotone objective function in (57), vertex k is fathomed. For $j = 1, \ldots, j(k) - 1$ it can be shown that $S_{k(j)}$ will be considered, if necessary, elsewhere in the tree. A possible tree is shown in Figure 5. Underlined vertices are assumed to be fathomed. Note that in branching from vertex 3, we do not consider $x_1 \geq 1$ since $S_3 \cap \{x_N | x_1 \geq 1\} = S_6$.

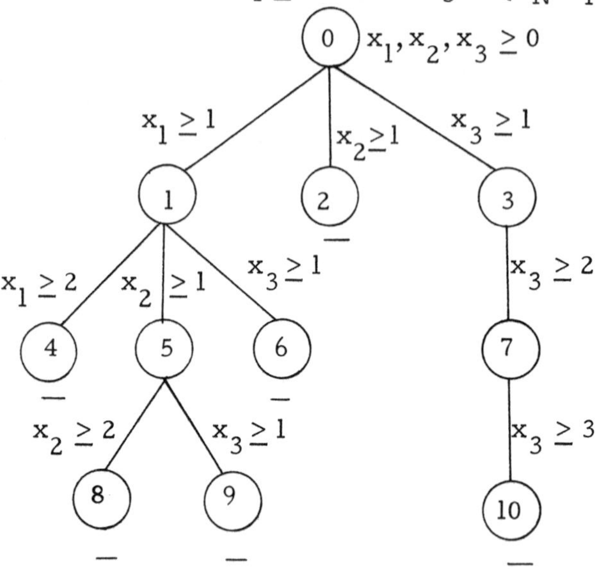

Figure 5

5. Cutting Plane Algorithms

Another class of algorithms for the ILP (2) operates by generating a set

$$T = \{x | Ax = b, \bar{A}x = \bar{b}, x \geq 0\}, \quad S \subseteq T \tag{59}$$

such that the relaxation of (2)

$$\max cx, \quad x \in T \tag{60}$$

has an optimal integer solution x^0. Thus x^0 is an optimal solution to (2). Clearly, if $S \neq \emptyset$, a set T as given by (59) exists, since the convex hull of S will suffice.

Suppose we have a basic representation of the LP corresponding to (2) given by

(61) $$x_{B_i} = y_{i0} - \sum_{j \in R} y_{ij} x_j, \quad i = 0, \ldots, m$$

which are the equations of (27) with $z = x_{B_0}$ and R the index set of nonbasic variables. Multiplying the ith row of (61) by $h \neq 0$ yields

$$hx_{B_i} + \sum_{j \in R} hy_{ij} x_j = hy_{i0}$$

and then $x \geq 0$ implies

(62) $$[h]x_{B_i} + \sum_{j \in R} [hy_{ij}]x_j \leq hy_{i0} .$$

Since x must be integer, the left hand side of (62) must be integer, and thus

(63) $$[h]x_{B_i} + \sum_{j \in R} [hy_{ij}]x_j \leq [hy_{i0}] .$$

Multiplying (61) by $[h]$ and subtracting (63) yields

(64) $$\sum_{j \in R} ([h]y_{ij} - [hy_{ij}])x_j \geq [h]y_{i0} - [hy_{i0}]$$

which is the fundamental <u>cut</u> derived from the <u>source</u> row i.

5.1 Method of Integer Forms [Gomory (1958), (1963a)]

In the method of integer forms we let h be an integer in (64). This yields the cut

(65) $$\sum_{j \in R} (hf_{ij} - [hf_{ij}])x_j \geq hf_{i0} - [hf_{i0}]$$

where f_{ij} is the fractional part of y_{ij}. For the purpose of illustration, let $h = 1$ in (65). This yields

(66) $$\sum_{j \in R} f_{ij} x_j \geq f_{i0} .$$

Clearly if $f_{i0} > 0$, the solution $x_j = 0$, $j \in R$ violates (66). Also, (66) does not exclude any point in S, since it has been derived solely from the assumptions of nonnegativity and integrality of the variables.

If (66) is written as the equation

(67) $$s = -f_{i0} + \sum_{j\in R} f_{ij}x_j, \quad s \geq 0$$

the new variable s is also integer, since from (61)

$$-s = x_{B_i} - [y_{i0}] + \sum_{j\in R} [y_{ij}]x_j \; .$$

Thus, (67) can be appended to a dual feasible LP solution (61), and the new problem can be solved by the dual simplex method. New cuts can be taken, in turn, from (67). If a solution is obtained that is integer and primal and dual feasible, it is optimal to the ILP (2).

In algorithmic form we have

Step 1: (Initialization). Solve the LP corresponding to (2). Go to Step 2.

Step 2: (Optimality test). Is the solution integer ? If so, it is optimal to (2). If not go to Step 3.

Step 3: (Cutting and pivoting). Choose a source row r with $f_{r0} > 0$ and add the cut (67) with $i = r$ to the bottom of the tableau. Reoptimize using the dual simplex algorithm. Go to Step 2.

Finiteness

With certain specializations, the algorithm above can be shown to be finite. These are:

a. In Step 3, use the lexicographic dual simplex algorithm so that lexicographically positive columns are maintained.

b. Delete the row corresponding to a slack variable from a cut if such a variable becomes basic.

c. Choose r to satisfy $r = \min\{i \mid f_{i0} > 0\}$.

One drawback of this algorithm is that there is no

upper bound on the number of iterations [Rubin (1970), Jeroslow and Kortanek (1971)]. The algorithm is also very sensitive to roundoff errors, since it is imperative to recognize integers correctly. Thus, one is motivated to develop cuts which produce integral tableaus.

5.2 Dual All-Integer Algorithm [Gomory (1963b)]

Consider h in (64) in the range $0 < h < 1$. Then (64) with $i = r$ is

(68) $$s = [hy_{r0}] - \sum_{j \in R} [hy_{rj}]x_j\,, \quad s \geq 0 \text{ integer.}$$

It can be shown that for some such h, an element $[hy_{rk}]$ in (68) of minus one can be obtained, which is the pivot element by the dual simplex pivot rules. Thus, the cut (68) may be appended, after every dual simplex iteration to maintain dual feasible, all-integer solutions. Of course if the ordinary pivot element is minus one, no cut is needed. To start the algorithm, however, it is necessary to construct an initial solution of this form, and one loses the advantage of operating in the neighborhood of the LP optimum.

Finiteness of the dual all-integer algorithm can also be shown with specializations similar to those required in the method of integer forms. As in the method of integer forms, however, no upper bound can be given on the number of iterations [Finkelstein (1970)]. Premature termination leaves the user with no feasible solution to the ILP. Thus, it would be desirable to be able to produce a sequence of feasible ILP solutions.

5.3 Primal All-Integer Algorithm [Young (1965), (1968), Glover (1968a)]

Consider a primal feasible, all-integer simplex tableau. If the ordinary primal simplex pivot element y_{rk} is 1, a simplex iteration will preserve integrality. Otherwise, let $h = \frac{1}{y_{rk}} < 1$ in (64) with $i = r$, where r would be the pivot row in the primal simplex algorithm. Then adjoin the cut

(69) $$s = \left[\frac{y_{r0}}{y_{rk}}\right] - \sum_{j\in R}\left[\frac{y_{rj}}{y_{rk}}\right]x_j\ , \quad s \geq 0 \text{ integer.}$$

Since $\left[\frac{y_{r0}}{y_{rk}}\right] \leq \frac{y_{r0}}{y_{rk}}$ we can pivot on x_k and s and obtain a pivot element of 1. Therefore, by beginning with a primal feasible integer tableau, we can maintain primal feasible integer solutions. The cut (69) does not exclude the current integer solution (it might be optimal); but unless y_{r0}/y_{rk} is an integer, the solution that would have been obtained by introducing x_k into the basis and removing x_{B_r} is eliminated. With appropriate specializations, finiteness can also be demonstrated for this algorithm. However the cut (69) is frequently degenerate $\left(\left[\frac{y_{r0}}{y_{rk}}\right] = 0\right)$ and one can expect to encounterlong (but finite after the necessary specializations) sequences of iterations with no change in the solution.

5.4 Cuts for the MILP [Beale (1958), Gomory (1960)]

Consider an optimal LP basic representation for an MILP whose i^{th} row is given by

(70) $$x_{B_i} = y_{i0} - \sum_{j\in R_1} y_{ij}x_j - \sum_{j\in R_2} y_{ij}v_j$$

where $R_1 \cup R_2 = R$. Assume that $f_{i0} > 0$ and that x_{B_i} is required to be integer. Taking fractional parts in (70) yields

(71) $$\sum_{j\in R_1} f_{ij}x_j + \sum_{j\in R_2} y_{ij}v_j - f_{i0} = [y_{i0}] - \sum_{j\in R_1} [y_{ij}]x_j - x_{B_i}.$$

Now, the right-hand side of (71) is integer, so the left-hand side must be as well. Thus, either

(72) $$\sum_{j\in R_1} f_{ij}x_j + \sum_{j\in R_2} y_{ij}v_j - f_{i0} \geq 0$$

or

(73) $$\sum_{j\in R_1} f_{ij}x_j + \sum_{j\in R_2} y_{ij}v_j - f_{i0} \leq -1 .$$

Let

$$R_2^+ = \{j \mid j \in R_2, \; y_{ij} > 0\} \text{ and } R_2^- = \{j \mid j \in R_2, \; y_{ij} < 0\}.$$

Then if (72) holds, so does

(74) $$\sum_{j\in R_1} f_{ij}x_j + \sum_{j\in R_2^+} y_{ij}v_j \geq f_{i0}$$

and if (73) then

(75) $$\sum_{j\in R_2^-} y_{ij}v_j \leq -1 + f_{i0} .$$

Multiplying (75) by $f_{i0}/(-1 + f_{i0}) < 0$ yields

(76) $$-\sum_{j\in R_2^-} \frac{f_{i0}y_{ij}v_j}{1 - f_{i0}} \geq f_{i0} .$$

The left-hand sides of (74) and (76) are both non-negative, and since (74) or (76) holds

(77) $$\sum_{j\in R_1} f_{ij}x_j + \sum_{j\in R_2^+} y_{ij}v_j - \sum_{j\in R_2^-} \frac{f_{i0}y_{ij}v_j}{1 - f_{i0}} \geq f_{i0}$$

which is a cut for an MILP. For an ILP, $R_2^+ = R_2^- = \phi$ and (77) specializes to (66). Note that when (77) is transformed into an equality, the surplus variable is not required to be integer.

By a similar argument the stronger cut

(78) $$\sum_{j\in R_1^+} f_{ij}x_j + \sum_{j\in R_1^-} \frac{f_{i0}(1-f_{ij})x_j}{1-f_{i0}} + \sum_{j\in R_2^+} y_{ij}v_j - \sum_{j\in R_2^-} \frac{f_{i0}y_{ij}v_j}{1 - f_{i0}} \geq f_{i0}$$

can be derived, where $R_1^+ = \{j \mid j \in R_1,\ f_{ij} \leq f_{i0}\}$ and $R_1^- = R_1 - R_1^+$. Finiteness of an algorithm based on the cuts (77) or (78) can be shown if and only if the objective function is constrained to be integer. Still stronger cuts for MILP's have been derived in Gomory and Johnson (1972) using results on corner polyhedra.

5.5 Intersection Cuts [Balas (1971), Balas et al. (1971)]

Another class of cuts, not based on (64), which have appealing geometric properties are known as intersection cuts. Essentially they are derived by considering the intersection of a closed, bounded convex set C containing y_0 (the optimal LP solution vector) but no integer points in its interior, and the convex cone determined by the extreme rays emanating from y_0 given by the nonbasic columns of the simplex tableau. The unique set of intersection points determine a hyperplane, which can be shown to be a valid cut. Such a cut is shown in Figure 6. One possible choice of C is a hypersphere.

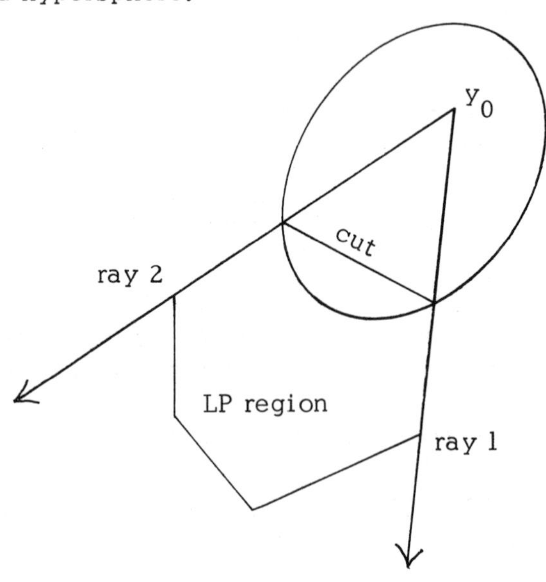

Figure 6

6. Approximate Methods [Echols and Cooper (1968), Senju and Toyoda (1968), Hillier (1969a), Mueller-Merbach (1970), Roth (1970)]

Many real problems are too large to be solved by exact algorithms. For these problems, clever approximate methods (heuristics) are a logical alternative. In addition heuristics provide cheap methods for getting good starting points for enumerative algorithms.

A neighborhood $N(x)$ of an integer x is a set of points in some sense "near" x. A neighborhood that is frequently used is the set of integer vectors that differ from x in at most a given number of components.

For the ILP (2) a point $x^0 \in S$ is said to be a local maximum with respect to the neighborhood $N(x)$ if

$$cx^0 \geq cx, \quad x \in S \cap N(x^0). \tag{79}$$

Most approximate methods generate a set of local optima. A general approximate method is:

Step 1: Select q neighborhoods, $N^1(x), \ldots, N^q(x)$ such that $N^t(x) \not\subset N^k(x)$, $t > k$. (The neighborhoods $N^2(x), \ldots, N^q(x)$ are used in an attempt to "escape" from local optima with respect to $N^1(x)$.) Choose $x^* \in S$ and let $\underline{z} = cx^*$. Go to Step 2.

Step 2: Solve

$$\max cx, \quad x \in S \cap N^1(x^*). \tag{80}$$

Go to Step 3.

Step 3: If (80) is unbounded, so is (2). If x^0 solves (80) and $cx^0 > cx^*$, let $x^* = x^0$ and go to Step 2. If $cx^0 = cx^*$, let $\underline{z} = \max\{cx^0, \underline{z}\}$ and $i = 2$. Go to Step 4.

Step 4: Attempt to find a point x' such that

$$(81) \qquad cx' > cx^*, \quad x' \in S \cap N^i(x^*) .$$

If a solution x' to (81) is found, let $x^* = x'$ and go to Step 2. Otherwise, if $i < q$, let $i = i+1$ and repeat Step 4. If $i = q$ go to Step 5.

Step 5: The solution that yielded $\underline{z}$ is the best known. Either choose another $x^* \in S$ and go to Step 2, or terminate.

The efficiency of the heuristic depends on the choice of neighborhoods, which affects the difficulty of solving (80) and (81). For some problems it may not be easy to find starting points in Step 1. In that case, a phase 1 heuristic can be developed to locate such points. Clearly, the local optimum obtained at each iteration depends on the starting point. Good starting points can often be obtained by examining lattice points near the LP optimum.

7. Computational Experience†

In this section we give an indication of the computational efficiency of various general and special purpose integer programming algorithms. The algorithms cited include those described in the rest of the paper and others. No attempt is made to describe algorithms not previously discussed.

Almost every author who reports computational results also invents his own parameters, which he feels are important in generating meaningful problems. Since there is no obvious consistency in these parameters, they have generally been omitted. In general, from the many data points listed in an article, a few are chosen in order to convey some typical computer runs and the maximum-size problem attempted.

†This section is based on Chapter 11 of Garfinkel and Nemhauser (1972b).

Table 3

	Variables	Constraints*
Haldi 1-6	5	4 (fixed charge)
Haldi 7-8	5	2 (fixed charge)
Haldi 9	6	6 (fixed charge)
Haldi 10	12	10 (fixed charge)
IBM 1, 2	7	7 (binary A matrix)
IBM 3	7	3
IBM 4, 5	15	15 (binary A matrix)
IBM 6	31	31 (binary A matrix)
IBM 7	50	12
IBM 8	37	12
IBM 9	35	15 (set covering)

*all are inequalities, nonnegativity not included.

An exception, where comparisons are possible, is a set of ILP's reported in Haldi (1964). A number of authors have attempted these problems, which are summarized in Table 3. Wherever possible, the results of these problems are reported. Unfortunately, the Haldi problems are rather small for making meaningful comparisons concerning many current algorithms. A new set of larger test problems is needed.

Another difficulty in reporting computational results is the number of different computers used. Every computer has its own characteristics, and it is not easy to make meaningful statements like "computer A is k times faster than computer B." Also, the skill of the programmer and efficiency of the programming language are very relevant but difficult to evaluate. We will simply list the computer used and let the reader draw his own conclusions.

Finally, there is a problem in reporting results on approximate methods. There is virtually no point in reporting times to find a suboptimal solution without also giving its deviation from optimality. Unfortunately, for the large problems for which approximate methods are designed, optimal solutions are generally unknown.

One source of our information has been the survey of Geoffrion and Marsten (1972), which contains a substantial amount of computational experience. In a few cases we cite results from Geoffrion and Marsten that are from unpublished sources to which we have not had access.

7.1 General Methods

Enumeration Algorithms

We use the notation (m, n) to describe an ILP with m constraints and n variables, and (m, n, p) for an MILP with m constraints, n integer variables and p continuous variables.

Roy, Benayoun, and Tergny (1970) have devised a commercial code using a branch and bound algorithm along the lines of the one described in Section 4.2. Some results for runs on the CDC 6600 are given in Table 4.

Table 4

m Constraints	p Continuous Variables	n Integer Variables	Time(sec)
288	244	59	27.9
68	0	536	6.6
604	1955	25	1146.0
1244	3884	24	361.6

A similar algorithm is described in Benichou et al. (1970). In terms of integer variables, the largest problem solved had (m, n) = (69, 590), and was solved in 25.7 minutes on an unspecified model of the IBM 360. An MILP with (m, n, p) = (721, 39, 1117) was solved in about 18 minutes.

Tomlin (1971) reports that use of penalties based on the cuts (78) can reduce running times in predominately integer problems by 50 percent or more over a similar algorithm not using these penalties. For instance, for an ILP with (m, n) = (5, 50) the running time on the UNIVAC 1108 was reduced from 93 to 42 seconds.

Computational experience with some very large problems is reported in Tomlin (1970). These include a binary MILP with (m, n, p) = (200, 40, 700) and two binary ILP's of size about (m, n) = (800, 200) and (1000, 200) respectively. Integer solutions within 3 percent of the continuous optima were obtained, although they could not be proved optimal. Computations were done on a UNIVAC 1108. The first problem took about 10 minutes, the latter two more than 45 minutes.

Davis, Kendrick, and Weitzman (1971) have also developed an algorithm of the same general type for binary MILP's. Running on the IBM 7094, the largest problem attempted had (m, n, p) = (197, 100, 217) and was solved in 5 minutes.

Davis (1969) has also developed an algorithm for the binary MILP. It is of a branch and bound variety, but also uses constraints from Benders' decomposition of MILP's into ILP's. Running on the UNIVAC 1108, the code solved all of the Haldi fixed charge problems in less than 1 second. Two scheduling problems with (m, n) = (40, 131) were solved within 20 seconds.

Aldrich (1969) has coded a version of an algorithm based on Benders' decomposition for the IBM 360/65. The ILP's are solved by a variation of the implicit enumeration algorithm. A problem with (m, n, p) = (39, 14, 53) was solved in 4.5 minutes. The largest problem solved had (m, n, p) = (378, 24, 136), but computation time was not reported.

Childress (1969) has used a similar algorithm to obtain good solutions to very large MILP's. A feasible solution within 1.3 percent of optimum was obtained on a problem with $(m, n, p) = (515, 48, 851)$ in about 20 minutes on a UNIVAC 1108. This type of method has also been used by Manne (1971), who reports solving a binary MILP with $(m, n, p) = (531, 22, 622)$ in about 30 minutes on an IBM 360/67.

Geoffrion (1969) reports contrasting experience with the basic implicit enumeration algorithm and one incorporating surrogate constraints via linear programming. The computer used was the IBM 7044. Table 5 contains some of the data points for general test problems, where upper bounds of 3 and 1 have been imposed on the variables for IBM 5 and 6, respectively. Table 6 gives results for set covering problems. Times in Table 6 are averages for five problems using surrogate constraints at every vertex. Seven percent of the entries in the A matrices are 1's (density = 0.07). Two 30-variable problems were solved without the surrogate constraints in 1.2 and 1.6 minutes; however, all attempted 40-variable problems exceeded their 16-minute time limit. Geoffrion and Marsten (1972) state that this code has undergone extensive modifications. It now runs much more efficiently and has the capability of solving much larger problems. A diverse set of eight binary ILP/s with $80 \leq n \leq 131$ were solved on the IBM 360/91. The computing times ranged from 0.95 to 7.5 seconds.

Trotter and Shetty (1971) have extended the surrogate constraint algorithm to handle general integer variables directly. Times on the CDC 6600 for the Haldi and IBM problems, treated directly and by binary expansion, are given in Table 7. These results indicate that direct treatment is more efficient.

Thangavelu and Shetty (1971) have used an implicit enumeration algorithm with surrogate constraints to solve assembly line balancing problems. The largest problem considered had $(m, n) = (73, 450)$ and was solved in less than 5 seconds on a UNIVAC 1108. Nine other problems were solved in less than 1 second. Additional computational experience with implicit enumeration algorithms on combinatorial problems is given by Ashour and Char (1971).

Table 5

Problem	m	n	Time (sec) No LP	LP
	5	39	>600	25.8
	5	50	>600	27.6
Haldi 8	4	20	6.0	3.0
10	10	30	24.6	3.6
IBM 1	7	21	8.4	0.6
5	15	30	>600	114.0
6	31	31	>600	120.0

Table 6

m	n	Time (sec)
30	30	1.8
30	40	4.2
30	50	4.8
30	60	8.4
30	70	9.0
30	80	12.6
30	90	10.2

Table 7

Problem	Time (sec)	
	Direct	Binary Expansion
Haldi 1	0.03	0.12
2	0.03	0.13
3	0.03	0.20
4	0.02	0.08
7	0.11	2.03
8	0.14	1.57
9	0.03	0.23
10	0.29	4.22
IBM 1	0.06	0.27
2	0.14	0.89
3	0.02	0.24
4	1.50	9.23
5	53.37	519.66
6	62.92	747.73
7	3.33	19.10
8	79.33	22.69
9	4.75	4.75

In Salkin (1970), the importance of a good starting point for implicit enumeration algorithms is illustrated. In fact, it is seen to be often advantageous to restart the enumeration process when a good feasible solution is discovered. The time sacrificed in possibly repeating some of the previous enumeration seems to be more than offset by the flexibility obtained by allowing those variables previously fixed at zero to be free. A good deal of computational experience is reported, but in a form difficult to summarize meaningfully.

Hillier (1969b) has derived another enumerative algorithm for ILP's. He uses an efficient heuristic procedure [see Hillier (1969a)] to generate good starting points. Computations are on the IBM 360/67, and the problems are randomly generated using various parameters. Two problems with $(m, n) = (30, 15)$ each took less than 4 seconds. Two others with $(m, n) = (15, 30)$ took 5 and 160 seconds, respectively. The two largest problems had $(m, n) = (30, 30)$, but neither terminated in 5 minutes.

Korte, Krelle, and Oberhoffer (1969), (1970) report results with an enumeration algorithm based on lexicographic search. They contrast their results with those of many other authors. All of the Haldi and IBM problems, except IBM 6 and 7, were attempted with times ranging from 0.5 seconds (Haldi 1-4, 9, IBM 5) to 484 seconds (IBM 8) on an IBM 7090.

Many authors have reported that enumerative algorithms often find an optimal solution quite early in the search so that the majority of the time is spent verifying optimality. Thus premature termination often results in a very good feasible, if not optimal, solution.

Cutting Plane Algorithms

Trauth and Woolsey (1969) report computational results that contrast Gomory's method of integer forms with his dual all-integer algorithm. A total of five codes based on the two algorithms are compared. Here we give results for two of them. The first, known as LIP1, is a version of the method of integer forms developed by Haldi and Issacson (1965). It is written in FORTRAN and FAP (IBM 7090 assembly language) and has a maximum problem size of $(m, n) = (60, 240)$. The second code, called ILP2, is based on the dual all-integer algorithm. Runs of LIP1 were on the IBM 7090, for ILP2 on the CDC 3600. An upper bound of 7000 was used on the number of iterations for both codes where iterations denote constraints added. Note that there will, in general, be many dual simplex iterations for each new constraint in LIP1, but only one in ILP2. The first set of problems attempted were ten variable binary knapsack problems. Some Haldi and IBM problems were also run. Results are in Table 8.

Srinivasan (1965) investigates various source row selection rules in the method of integer forms. He considers 8 ILP's with $11 \leq m \leq 48$ and $14 \leq n \leq 98$ and various combinations of selection rules too complex to describe here. Computations were done on the IBM 7094. The best rule varied from problem to problem. One of the more difficult of the eight problems with $(m, n) = (37, 44)$, was solved in about 32 seconds and required 289 pivots using the best rule for it. To contrast the different rules, a problem with $(m, n) = (10, 20)$ was solved in about 4 seconds and 71 pivots using its best rule, but could not be solved in 600 pivots using other reasonable rules. Finally, another group of 6 ILP's that could not be solved with any of the rules in 600-900 pivots is given. One of these problems is as small as $(m, n) = (10, 7)$.

Geoffrion and Marsten (1972) state that Martin reports success in solving a number of large set covering problems on the CDC 3600 using his accelerated euclidean algorithm [Martin (1963)]. This algorithm combines cuts in the method of integer forms in an attempt to achieve integer solutions more rapidly. Crew scheduling problems with $(m, n) = (100, 4000)$ typically take about 10 minutes; larger ones with $(m, n) = (150, 7000)$ are often solved using only a few cuts in 40 minutes.

Other authors also report that set covering and partitioning problems often require very few cuts after an optimal solution to the corresponding LP has been obtained. Of the 10 partitioning problems, ranging in size from $(m, n) = (5, 30)$

Table 8

Problem		LIP 1		ILP 2	
		Time (sec)	Iterations	Time (sec)	Iterations
Knapsack	b = 35	2.4	19	1.9	51
	b = 70	2.6	19	1.9	48
	b = 90	4.5	51	2.0	57
	b = 100	1.9	12	1.6	34
Haldi	6	7.6	123	3.3	311
	7	7.8	159		>7000
	8	6.4	126	3.0	306
	9	3.2	42	3.6	298
	10	9.2	102		>7000
IBM	1	1.9	11	1.1	11
	2	3.0	32	1.1	15
	3	2.9	53	0.6	14
	4	11.7	73	3.1	18
	5	66.4	351	26.2	842
	9	473.1	953	75.1	1105

to $(m, n) = (15, 388)$, considered by Balinski and Quandt (1964), only 3 required more than two cuts and only 1 required more than seven cuts.

Toregas, et al. (1971) have solved more than 150 structured set covering problems arising from location applications. These problems have square A matrices, unit costs, and as many as 50 constraints. In most cases no cuts were required and more than one cut was never needed.

Baugh, Ibaraki, and Muroga (1971) report solving problems as large as $(m, n) = (91, 240)$, using the dual all-integer algorithm. Computation times are not given.

Almost no computational experience has been obtained for the primal all-integer algorithm. However, Arnold and Bellmore (1971) have attempted 20 randomly generated problems having $(m, n) = (10, 10)$. Fourteen of those problems required 58 or fewer iterations. However, none of the others was solved in 500 iterations.

ILPC-Based Algorithms

Shapiro (1968b), and Gorry and Shapiro (1971) report computational results for the branch and bound algorithm that uses the ILPC relaxation. The algorithm of the 1971 paper uses a different separation in the branch and bound phase than the one given in Section 4.4 and also finds all alternative optima to the ILPC. Nine of the 17 problems reported in the 1968 paper were solved without enumeration. Three others required enumeration only because all alternative optima to the ILPC were not generated before the branch and bound phase. Of 25 problems considered in the 1971 paper, 19 did not require enumeration. On the negative side it is indicated that some ILPC's could not be solved because of very large determinants. Bradley and Wahi (1969) solve ILPC's for 28 test problems using a transformation to Hermite Normal Form. In 23 of the 28 problems some optimal solution to the ILPC was feasible to the ILP. Although there was some overlap in the problem sets considered by Shapiro, Gorry and Shapiro, and Bradley and Wahi, in each case about 75 percent of the ILP's could be solved as ILPC's.

Some results in Shapiro (1968), on the Haldi and IBM problems are shown in Table 9 . The computer is the IBM 360/65 and D is the absolute value of the determinant of the optimal basis matrix. For some of these problems, slightly better times are reported in Gorry and Shapiro.

Table 9

Problem	D	Time (sec)	ILPC Solves ILP
Haldi 1	183	1.3	Yes
2	258	1.0	Yes
3	320	1.3	Yes
4	205	0.9	Yes
9	2000	6.6	Yes
IBM 1	32	2.2	Yes
2	32	2.0	Yes
3	72	0.8	No
8	2856	9.5	Yes
9	96	20.2	No

Some of the larger problems considered in Gorry and Shapiro were solved on a UNIVAC 1108. Included in these were a group of media selection problems having $(m, n) = (75, 150)$, which were solved in under a minute. The largest problem considered was a crew scheduling problem with $(m, n) = (104, 236)$; it was solved in about 7 minutes. Gorry and Shapiro also indicate that cutting planes and other procedures for reducing the size of determinants have been incorporated into their algorithm so there is a greater possibility of solving ILPC's when the optimal basis matrix has a very large determinant [see also, Gorry, Shapiro, and Wolsey (1972)].

Approximate Methods

Roth (1970) gives computational experience for a variation of the approximate method described in Section 6. Some results for standard problems are given in Table 10. Times specify the average for 25 iterations and f indicates the proportion of iterations in which the optimal solution was found. Runs are on the GE 635. The parameter k indicates the number of components of x that can differ from x^0 in the definition of N(x) .

Table 10

Problem	k = 5		k = 3	
	f	time(sec)	f	time (sec)
IBM 4	0.92	8.2	0.88	0.9
6	0.8	360.0	0.12	8.0
9	0.4	5.0	0.32	0.5

Echols and Cooper (1968) give a direct search approximate method for the general ILP, which is also similar to the approximate method of Section 6. Computations were done on the IBM 7072, with problems ranging up to (m, n) = (27,21). For these small problems the optimal solution was found in almost every case. Times ranged from 0.006 to 9.3 minutes.

Hillier (1969a) reports results for another direct search approximate method. Solving problems of size (m, n) = (15, 15) and (30, 15), using various points near the LP optimum as starting points, he is generally able to find an optimal solution. Times are in the order of a few seconds on the IBM 360/67.

7.2 Special Purpose Methods

Knapsack Problem

Cabot (1970) devised an enumerative algorithm for the knapsack problem which is based on Fourier-Motzkin elimination. The a_j's and c_j's were generated from the uniform distribution between 10 and 110, and $b_j = 2\Sigma_{j=1}^{n} a_j$. The computer was the CDC 3600. Results are given in Table 11. For each value of n , 50 problems were run.

Table 11

n	Average time (sec)	Maximum time (sec)
60	0.41	3.5
70	0.66	3.5
80	0.86	7.2

Greenberg and Hegerich (1970) use a branch and bound algorithm to solve the binary knapsack problem. The a_j's and c_j's are randomly generated (the distribution used is not reported). The computer is the IBM 360/67. Problems with $b = 100$ and $n = 50$ took, on the average, 0.163 seconds. It is also reported that several problems with 5000 variables were solved in approximately 4 minutes.

Nemhauser and Ullmann (1969) give results for an algorithm based on dynamic programming. The variables are bounded. The largest run had a_j's and c_j's uniformly random between 0 and 99, $b = 2500$, $n = 50$, and $x_j \leq 2$ for all j . Computation time was 91.2 seconds on the IBM 7094.

Unfortunately, we cannot summarize any computational experience with the Gilmore and Gomory [(1961), (1963), (1965), and (1966)] algorithms. These papers contain substantial experience, but the data are based on cutting stock problems so that it is difficult to extract the data and times for the knapsack problems.

Edge Packing

Gordon (1971) has coded versions of the Balinski (1969) and Edmonds (1965a) algorithms for finding maximum cardinality edge matchings on graphs. His results indicate slightly faster running times for Balinski's algorithm. The largest problem considered has 60 vertices and 263 edges. It was solved in less than 2 seconds on an IBM 360/65 with both methods.

Edmonds and Johnson (1970) report results for their algorithm for b-weighted edge matching problems. A typical run for a problem with 300 vertices, 1500 edges, $b_i = 1, 2$, and $1 \leq c_j \leq 10$ takes about 30 seconds on an IBM 360/91.

Set Covering and Partitioning

Bellmore and Ratliff (1971) have developed a cutting plane algorithm and have coded it for the IBM 7094. For the problems of Table 12, costs are random between 1 and 99. Failures to solve problems were caused by the core of limitations of the computer.

An implicit enumeration algorithm was run on the IBM 360/50 by Lemke, Salkin, and Spielberg (1971). Their computational results are hard to summarize, since a large number of parameters are introduced. One run with $(m, n) =$ $(30, 90)$ and density of 0.07 took 36 seconds, and another with $(m, n) = (50, 450)$ and density of 0.046 took 2.15 minutes.

Table 12

m	n	Density	Problems Attempted	Problems Solved	Average Time (sec)
50	80	0.07	10	9	11.5
50	90	0.07	10	8	19.7
30	80	0.21	10	9	8.5
30	90	0.21	10	10	7.6

An implicit enumeration for the set partitioning problem has been coded in FORTRAN and MAP (IBM 7094 assembly language) by Garfinkel and Nemhauser (1969). Some results are given in Table 13. The first 15 problems were randomly generated, while the sixteenth was adapted from a redistricting problem. As illustrated by problems 10 and 11, high density problems are easier to solve than low density ones.

Table 13

Problem	m	n	Density	Time (sec)
1-9	20	100	0.1-0.5	2-5
10	37	200	0.3	5
11	37	200	0.1	257
12	26	385	0.16	25
13	26	777	0.16	39
14	100	1400	0.15	895
15	37	1790	0.25	>900
16	37	1790	0.25	49

A similar algorithm has also been run by Pierce (1968) with comparable results. Pierce and Lasky (1970) have incorporated surrogate constraints, linear programming and other modifications into the algorithm and have obtained some better results. Marsten (1971), using a different scheme, enumerates on the rows rather than the columns and also uses linear programming. His algorithm is considerably more efficient for low density problems, but does not perform as well on high density ones. Marsten has solved several large crew scheduling problems including one with $(m, n) = (117, 4845)$, in 8 minutes on an IBM 360/91. A still larger one with $(m, n) = (400, 22000)$ could not be solved. In fact the corresponding LP could not be solved in 3 hours.

Fixed Charge Problem

Steinberg (1970) has developed several approximate methods, as well as a branch and bound algorithm for the general fixed charge problem. The algorithm, run on the IBM 7072, solved binary MILP's with $(m, n, p) = (15, 30, 30)$ in from 4.8 to 47.3 minutes. For the same size problems, however, one of the heuristics was generally able to get within 3 percent of the optimal solution.

Spielberg (1969a) has developed an implicit enumeration algorithm that uses constraints from Benders' decomposition and has applied it to facilities location problems. His model is a binary MILP with m being the number of plants plus the number of customers, n the number of plants, and $p = mn$. He gives a multitude of results that are difficult to summarize. Two of the larger problems had $(m, n, p) = (250, 100, 15000)$ and $(m, n, p) = (184, 92, 8464)$. They were solved in 30 and 46 minutes respectively on the IBM 360/50.

8. Summary and Synthesis

Developments in the use, computational aspects, and theory of integer programming have been substantial. There is a rich variety of integer programming models of real problems, and some measure of success has been achieved in solving them. However, problem size is still a major limitation.

For general MILP's, branch and bound programs are most widely used and have been most successful. Less substantial, but comparable, experience has been obtained using algorithms based on Benders' decomposition of MILP's into ILP's. The number of integer variables appears to be the most significant measure of difficulty using these approaches. Experience with commercial codes indicates that MILP's having fewer than 50 integer variables are extremely likely to be solvable, problems with 50-100 integer variables have a good change of being solved, and good feasible solutions can be obtained for substantially larger problems. All of these commercial codes have highly sophisticated linear programming subroutines. The limitation on the number of constraints and continuous variables depends heavily on the linear programming code. Successful runs have been obtained for problems having more than 500 continuous variables and constraints. Nevertheless it is still possible to encounter quite small problems that defy solution, since no reasonable upper bound on the number of calculations is known.

The use of both linear programming and some form of enumeration appears to be necessary to achieve good computational results for general ILP's and MILP's. Algorithms based exclusively on cutting planes or implicit enumeration without surrogate constraints (derived from optimal linear programming solutions) have not, in general, been effective for medium and large problems.

Although pure cutting plane algorithms appear to be of limited practical value, cuts can be very important in branch and bound algorithms. It has already been shown that improved penalties can be obtained from cuts. Moreover, cuts can be incorporated into a branch and bound algorithm in a more direct way. In the process of solving an ILP or MILP by branch and bound, an alternative to making the separation (49) is to use row i as a source row for a cut of the form (78). If there is no dual degeneracy, progress in reducing the objective function can be achieved without resorting to separation, thereby eliminating the resulting growth in tree size. Presumably, empirical rules could be established for choosing between these two options.

Another option is available for ILP's. If an optimal solution to an LP is not all-integer and the determinant of the basis matrix is not too large, a better bound and possibly an optimal ILP solution can be obtained by solving the corresponding ILPC. The point is that the various methods described are compatible and could all be subroutines in one master algorithm. At each iteration a supervisor routine or a human, if an interactive system is being used, could intervene and choose a tactic based upon the current status of the problem. [See Weingartner and Ness (1967), Gorry and Shapiro (1971), and Geoffrion and Marsten (1972).] Although an intelligent choice might be difficult and might require considerable adaptive learning, a plausible start along these lines could be as follows:

1. If the number of variables required to be integer, but which are not integer in the current solution, is less than K_1, where K_1 is a suitably small specified constant, separation is a good strategy. In this case the enumeration from the current solution should not be excessive.

2. If the problem is an ILP and the magnitude of the determinant of the current basis matrix is no greater than K_2, where K_2 is a suitably small specified constant, solving the corresponding ILPC is a good strategy. In this case the ILPC will be solved efficiently. If no optimal solution to the ILPC solves the ILP, we could enumerate as in Section 4.4 or just use the ILPC solution as an upper bound. The decision regarding enumeration could be based on the "amount" of infeasibility in the ILPC solution. Note that even if the current problem is an MILP, an ILPC can be derived by setting all of the nonbasic continuous variables to zero. If the number of continuous nonbasic variables is small, this approach might be used to find a good feasible solution to the MILP.

3. The option of adding a cutting plane and reoptimizing the LP is always available. This option is the simplest of the three in the sense that it requires much less work than (2) and does not cause an increase in the number of problems to be solved, as does (1). If a cut produces a satisfactory decrease in the dual objective function, adding it should be a good strategy.

Of course, it is to be expected that situations will be encountered in which none of the three strategies will appear to be attractive. However, there are other alternatives:

4. Generate a surrogate constraint in the hope that new information will become available; for example, a good source row for a cut.

5. Use an approximate method to obtain a good feasible solution.

The synthesis that we have proposed is based entirely on existing methodology -- advances in the theory of integer programming and computer technology will surely lead to quite different algorithms.

General ILP and MILP algorithms cannot be expected to perform as well on highly structured problems as special algorithms. A notable case is edge packing for which there is a good algorithm.

Special purpose algorithms have also markedly outperformed general algorithms for more difficult combinatorial optimization problems associated with graphs. A good illustration is the traveling salesman problem, where branch and bound schemes based on assignment, matching, and spanning tree relaxations have solved large problems. [Bellmore and Malone (1971), Held and Karp (1970), (1971)].

Structure in MILP's can be capitalized on, using the partitioning scheme of Benders, since the decomposition of an MILP into a sequence of ILP's and LP's preserves the structure of the continuous part of the problem. For example, in fixed charge transportation problems, the network flow structure of the LP is preserved.

There are numerous special purpose algorithms for set covering and partitioning problems. Based on factors such as the efficient handling of binary data, the ease of finding basic feasible integer solutions for the covering problem, these algorithms have been somewhat more successful than general algorithms. However, the solution of many real covering and partitioning models is beyond our current capabilities because the problems involve a very large number of variables and/or constraints. The development of efficient special purpose algorithms for these problems awaits a better understanding of the properties of structured binary matrices. [Fulkerson (1971), Berge (1972)].

Acknowledgment

We are grateful to Les Trotter for many helpful suggestions.

References

Abadie, J., ed. (1970), Integer and Nonlinear Programming. American Elsevier.

Agin, N. (1966), "Optimum Seeking with Branch-and-Bound," Man. Sci. 13, B176-B185.

Aldrich, D. W. (1969), "A Decomposition Approach to the Mixed Integer Programming Problem," Ph. D. Dissertation, School of Industrial Engineering, Purdue University.

Arnold, L. R. and M. Bellmore (1971), "Iteration Skipping in Primal Integer Programming," Dept. of Operations Research, The Johns Hopkins University.

Aronofsky, J., ed. (1969), Progress in Operations Research, Vol. 3: Relationship between Operations Research and the Computer, John Wiley & Sons.

Ashour, S. and A. R. Char (1971), "Computational Experience on 0-1 Programming Approaches to Various Combinatorial Problems," J. Op. Res. Soc. of Japan 13, 78-107.

Balas, E. (1965), "An Additive Algorithm for Solving Linear Programs with Zero-One Variables," Opns. Res. 13, 517-546.

Balas, E. (1967), "Discrete Programming by the Filter Method," Opns. Res. 15, 915-957.

Balas, E. (1971), "Intersection Cuts- A New Type of Cutting Planes for Integer Programming," Opns. Res. 19, 19-39.

Balas, E., V. J. Bowman, F. Glover, and D. Sommer (1971), "An Intersection Cut from the Dual of the Unit Hypercube," Opns. Res. 19, 40-44.

Balinski, M. L. (1965), "Integer Programming: Methods, Uses, Computation," Man. Sci. 12, 253-313.

Balinski, M. L. (1969), "Labelling to Obtain a Maximum Matching," Combinatorial Mathematics and its Applications, R. C. Bose and T. A. Dowling (eds.), University of North Carolina Press, 585-602.

Balinski, M. L. (1970), "On Maximum Matching, Minimum Covering and their Connections," 303-312, in Kuhn (1970).

Balinski, M. L. and R. E. Quandt (1964), "On an Integer Program for a Delivery Problem," Opns. Res. 12, 300-304.

Balinski, M. L. and K. Spielberg (1969), "Methods for Integer Programming: Algebraic, Combinatorial and Enumerative," 195-292, in Aronofsky (1969).

Baugh, C. R., T. Ibaraki, and S. Muroga (1971), "Results in Using Gomory's All-Integer Algorithm to Design Optimum Logic Networks," Opns. Res. 19, 1090-1096.

Beale, E. M. L. (1958), "A Method for Solving Linear Programming Problems When Some But Not All of the Variables Must Take Integral Values," Stat. Tech. Res. Group Tech. Rep. No. 19, Princeton University.

Bellmore, M. and J. C. Malone (1971), "Pathology of Traveling-Salesman Subtour Elimination Algorithms," Opns. Res. 19, 278-307.

Bellmore, M. and H. D. Ratliff (1971), "Set Covering and Involutory Bases," Man. Sci. 18, 194-206.

Benders, J. F. (1962), "Partitioning Procedures for Solving Mixed-Variables Programming Problems," Numerische Mathematik 4, 238-252.

Benichou, M., J. M. Gauthier, P. Girodet, G. Hentges, G. Ribiere, and O. Vincent (1971), "Experiments in Mixed-Integer Linear Programming," Math. Prog. 1, 76-94.

Berge, C. (1972), "Balanced Matrices," Math. Prog. 2, 19-31.

Bertier, P. and B. Roy (1964), "Procedure de Resolution pour une Classe de Problems Pouvant Avoir un Caractere Combinatore," Cahiers Cent. d'Etudes Recherche Operationelle 6, 202-208. Translated by W. Jewell (1967), ORC Rep. 17-34, University of California, Berkeley.

Bessiere, F. (1965), "Sur la Recherche du Nombre Chromatique d'un Graphe par un Programme Lineaire en Nombres Entiers," Rev. Franc. Recherche Operationelle 9, 143-148.

Bradley, G. H. (1971), "Transformation of Integer Programs to Knapsack Problems," Discrete Math. 1, 29-45.

Bradley, G. H. and P. N. Wahi (1969), "An Algorithm for Integer Linear Programming: A Combined Algebraic and Enumeration Approach," Rep. No. 29, Dept. of Administrative Science, Yale University.

Cabot, V. A. (1970), "An Enumeration Algorithm for Knapsack Problems," Opns. Res. 18, 306-311.

Childress, J. P. (1969), "Five Petrochemical Industry Applications of Mixed Integer Programming," Bonner and Moore Assoc. Inc., Houston, Texas.

Cook, S. (1971), "The Complexity of Theorem-Proving Procedures," ACM Conference on Theory of Computation, 151-158.

Dakin, R. J. (1965), "A Tree-Search Algorithm for Mixed Integer Programming Problems," Computer J. 8, 250-255.

Davis, R. E. (1969), "A Simplex-Search Algorithm for Solving 0-1 Mixed Integer Programs," Tech. Rep. No. 5, Dept. of Operations Research, Stanford University.

Davis, R. E., D. A. Kendrick, and M. Weitzman (1971), "A Branch-and-Bound Algorithm for 0-1 Mixed Integer Programming Problems," Opns. Res. 19, 1036-1044.

Dijkstra, E. W. (1959), "A Note on Two Problems in Connexion with Graphs," Numer. Mathematik 1, 269-271.

Echols, R. E. and L. Cooper (1968), "Solution of Integer Linear Programming Problems by Direct Search," JACM 15, 75-84.

Edmonds, J. (1962), "Covers and Packings in a Family of Sets," Bull. Am. Math. Soc. 68, 494-499.

Edmonds, J. (1965a), "Paths, Trees and Flowers," Can. J. Math. 17, 449-467.

Edmonds, J. (1965b), "Maximum Matching and a Polyhedron with 0,1-Vertices," J. Res. Nat. Bur. Stds. 69B, 125-130.

Edmonds, J. and E. L. Johnson (1970), "Matching: A Well-Solved Class of Integer Linear Programs," Proc. of the Calgary Int. Conf. on Comb. Structures and Their Appl., 89-92, Gordon and Breach.

Edmonds, J. and R. M. Karp (1972), "Theoretical Improvements in Algorithmic Efficiency for Network Flow Problems," JACM 19, 248-264.

Elmaghraby, S. E., and M. K. Wig (1970), "On the Treatment of Cutting Stock Problems as Diophantine Programs," Dept. of Industrial Engineering, North Carolina State University.

Finkelstein, J. J. (1970), "Estimation of the Number of Iterations for Gomory's All-Integer Algorithm," Dokl. Akad. Nauk, SSSR. 193, 988-992.

Fleischmann, B. (1967), "Computational Experience with the Algorithm of Balas," Opns. Res. 15, 153-155.

Ford, L. R., Jr. and D. R. Fulkerson (1962), Flows in Networks, Princeton University Press.

Fulkerson, D. R. (1971), "Blocking and Anti-Blocking Pairs of Polyhedra," Math. Prog. 1, 168-194.

Garfinkel, R. S. and G. L. Nemhauser (1969), "The Set Partitioning Problem: Set Covering with Equality Constraints," Opns. Res. 17, 848-856.

Garfinkel, R. S. and G. L. Nemhauser (1970), "Optimal Political Districting by Implicit Enumeration Techniques," Man. Sci. 16, B495-B508.

Garfinkel, R. S. and G. L. Nemhauser (1972a), "Optimal Set Covering: A Survey," 164-183, in Geoffrion (1972).

Garfinkel, R. S. and G. L. Nemhauser (1972b), Integer Programming, John Wiley & Sons.

Geoffrion, A. M. (1967), "Integer Programming by Implicit Enumeration and Balas' Method," SIAM Rev. 7, 178-190.

Geoffrion, A. M. (1969), "An Improved Implicit Enumeration Approach for Integer Programming," Opns. Res. 17, 437-454.

Geoffrion, A. M., ed. (1972), Perspectives on Optimization: A Collection of Expository Articles, Addison-Wesley.

Geoffrion, A. M. and R. E. Marsten (1972), "Integer Programming: A Framework and State-of-the-Art Survey," Man. Sci. 18, 465-491. Also 137-163 in Geoffrion (1972).

Gilmore, P. C. and R. E. Gomory (1961), "A Linear Programming Approach to the Cutting Stock Problem," Opns. Res. 9, 849-859.

Gilmore and R. E. Gomory (1963), "A Linear Programming Approach to the Cutting Stock Problem, Part II," Opns. Res. 11, 863-888.

Gilmore, P. C. and R. E. Gomory (1965), "Multistage Cutting Stock Problems of Two and More Dimensions," Opns. Res. 13, 94-120.

Gilmore and R. E. Gomory (1966), "The Theory and Computation of Knapsack Functions," Opns. Res. 14, 1045-1074.

Glover, F. (1965), "A Multiphase-Dual Algorithm for the Zero-One Integer Programming Problem," Opns. Res. 13, 879-919 .

Glover, F. (1968a), "A New Foundation for a Simplified Primal Integer Programming Algorithm," Opns. Res. 16, 727-740.

Glover, F. (1968b), "Surrogate Constraints," Opns. Res. 16, 741-749.

Glover, F. and R. E. Woolsey (1970), "Aggregating Diophantine Equations," Rep. No. 70-4, University of Colorado.

Glover, F. and S. Zionts (1965), "A Note on the Additive Algorithm of Balas," Opns. Res. 13 , 546-549.

Gomory, R. E. (1958), "Outline of an Algorithm for Integer Solutions to Linear Programs," Bull. Amer. Math. Soc. 64, 275-278.

Gomory, R. E. (1960), "An Algorithm for the Mixed Integer Problem," RM-2597, RAND Corp.

Gomory, R. E. (1963a), "An Algorithm for Integer Solutions to Linear Programs," 269-302, in Graves and Wolfe (1963). (Originally appeared as Princeton-IBM Math. Res. Project Tech. Rep. No. 1, 1958.)

Gomory, R. E. (1963b), "All-Integer Integer Programming Algorithm," 193-206, in Muth and Thompson (1963).

Gomory, R. E. (1965), "On the Relation between Integer and Non-Integer Solutions to Linear Programs," Proc. Nat. Acad. Sci. 53 , 260-265.

Gomory, R. E. (1967), "Faces of an Integer Polyhedron," Proc. Nat. Acad. Sci. 57 , 16-18.

Gomory, R. E. (1969), "Some Polyhedra Related to Combinatorial Problems," Lin. Alg. and Appl. 2, 451-558.

Gomory, R. E. and E. L. Johnson (1972), "Some Continuous Functions Related to Corner Polyhedra," Math. Prog. 3, 23-85.

Gordon, B. E. (1971), "The Maximum Matching Problem- A Comparison of the Edmonds and Balinski Algorithms," Graduate School of Management, University of Rochester.

Gorry, G. A. and J. F. Shapiro (1971), "An Adaptive Group Theoretic Algorithm for Integer Programming Problems," Man. Sci. 17, 285-306.

Gorry, G. A., J. F. Shapiro, and L. A. Wolsey (1972), "Relaxation Methods for Pure and Mixed Integer Programming Problems," Man. Sci. 18, 229-239.

Graves, R. L. and P. Wolfe, eds. (1963), Recent Advances in in Mathematical Programming, McGraw-Hill.

Greenberg, H. (1971), Integer Programming, Academic Press.

Greenberg, H. and R. L. Hegerich (1970), "A Branch Search Algorithm for the Knapsack Problem," Man. Sci. 16, 327-332.

Haldi, J. (1964), "25 Integer Programming Test Problems," Working Paper No. 43, Graduate School of Business, Stanford University.

Haldi, J. and L. M. Isaacson (1965), "A Computer Code for Integer Solutions to Linear Programs " Opns. Res. 13, 946-959.

Held, M., and R. M. Karp (1970), "The Traveling-Salesman Problem and Minimum Spanning Trees," Opns. Res. 18, 1138-1162.

Held, M., and R. M. Karp (1971), "The Traveling-Salesman Problem and Minimum Spanning Trees: Part II," Math. Prog. 1, 6-25.

Hillier, F. S. (1969a), "Efficient Heuristic Procedures for Integer Linear Programming with an Interior," Opns. Res. 17, 600-637.

Hillier, F. S. (1969b), "A Bound-and-Scan Algorithm for Pure Integer Linear Programming with General Variables," Opns. Res. 17, 638-679.

Hoffman, A. J. and J. B. Kruskal (1958), "Integral Boundary Points of Convex Polyhedra," 223-246, in Kuhn and Tucker (1958).

Hu, T. C. (1969), Integer Programming and Network Flows, Addison-Wesley.

Hu, T. C. (1970), "On the Asymptotic Integer Algorithm," Lin. Alg. and Appl. 3, 279-294.

Jeroslow, R. G. and K. O. Kortanek (1971), "On an Algorithm of Gomory," SIAM J. 21, 55-60.

Karp, R. M. (1972), "Reducibility Among Combinatorial Problems," Tech. Rept. 3, Computer Science, University of California, Berkeley.

Klee, V. and G. J. Minty (1970), "How Good is the Simplex Algorithm," Math. Note. No. 643, Math. Res. Lab., Boeing Sci. Res. Lab.

Korte, B., W. Krelle, and W. Oberhoffer (1969), "Ein Lexikographischer Suchalgorithmus zur Losung Allgemeiner Ganzzahliger Programmierungsaufgaben I, II," Unternehmensforschung 13, 73-98, 171-192.

Korte, B., W. Krelle, and W. Oberhoffer (1970), "Ein Lexikographischer Suchalgorithmus zur Losung Allgemeiner Ganzzahliger Programmierungsaufgaben - Nachtrag," Unternehmensforchung 14, 228-234.

Kuhn, H. W., ed. (1970), Proceedings of the Princeton Symposium on Mathematical Programming, Princeton University Press.

Kuhn, H. W. and A. W. Tucker, eds. (1958), Linear Inequalities and Related Systems, Princeton University Press.

Land, A. H. and A. G. Doig (1960), "An Automatic Method for solving Discrete Programming Problems," Econometrica 28, 497-520.

Lawler, E. L. (1971), "The Complexity of Combinatorial Computations: A Survey," Proc. Symp. on Computers and Automata.

Lawler, E. L. and D. E. Wood (1966), "Branch-and-Bound Methods: A Survey," Opns. Res. 14, 699-719.

Lemke, C. E., H. M. Salkin, and K. Spielberg (1971), "Set Covering by Single Branch Enumeration with Linear Programming Subproblems," Opns. Res. 19, 998-1022.

Lemke, C. E. and K. Spielberg (1967), "Direct Search Zero-One and Mixed Integer Programming," Opns. Res. 15, 892-914.

MacDuffee, C. C. (1940), An Introduction to Abstract Algebra, John Wiley & Sons.

Manne, A. S. (1971), "A Mixed Integer Algorithm for Project Evaluation," Int. Bank. in Reconstruction and Development, Washington, D. C.

Marsten, R. E. (1971), "An Implicit Enumeration Algorithm for the Set Partitioning Problem with Side Constraints," Ph. D. Dissertation, University of California, Los Angeles.

Martin, G. T. (1963), "An Accelerated Euclidean Algorithm for Integer Linear Programming," 311-318, in Graves and Wolfe (1963).

Mitten, L. G. (1970), "Branch-and-Bound Methods: General Formulation and Properties," Opns. Res. 18, 24-34.

Mueller-Merbach, H. (1970), "Approximation Methods for Integer Programming," Johannes Gutenberg University, Mainz, Germany.

Muth, J. F. and G. L. Thompson, eds. (1963), Industrial Scheduling, Prentice-Hall.

Nemhauser, G. L. and Z. Ullmann(1969), "Discrete Dynamic Programming and Capital Allocation," Man. Sci. 15, 494-505.

Padberg, M. (1970), "Equivalent Knapsack-type Formulations of Bounded Integer Linear Programs," Man Sci. Res. Rep. No. 227, Carnegie-Mellon University.

Petersen, C. C. (1967), "Computational Experience with Variants of the Balas Algorithm Applied to the Selection of R and D Projects," Man. Sci. 13, 736-750.

Pierce, J. F. (1968), "Application of Combinatorial Programming to a Class of All-Zero-One Integer Programming Problems," Man. Sci. 15, 191-209.

Pierce, J. F. and J. S. Lasky (1970), "Improved Combinatorial Programming Algorithms for a Class of All Zero-one Integer Programming Problems," IBM Cambridge Sci. Cent. Rep.

Roth, R. H. (1970), "An Approach to Solving Linear Discrete Optimization Problems," JACM 17, 303-313.

Roy, B., R. Benayoun, and J. Tergny (1970), "From S. E. P. Procedure to the Mixed Ophelie Program," 419-436, in Abadie (1970).

Rubin, D. S. (1970), "On the Unlimited Number of Faces in Integer Hulls of Linear Programs with a Single Constraint," Opns. Res. 18, 940-946.

Salkin, H. M. (1970), "On the Merit of the Generalized Origin and Restarts in Implicit Enumeration," Opns. Res. 18, 549-554.

Senju, S. and Y. Toyoda (1968), "An Approach to Linear Programming with 0-1 Variables," Man. Sci. 15, B196-B207.

Shapiro, J. F. (1968a), "Dynamic Programming Algorithms for the Integer Programming Problem - I: The Integer Programming Problem Viewed as a Knapsack Type Problem," Opns. Res. 16, 103-121.

Shapiro, J. F. (1968b), "Group Theoretic Algorithms for the Integer Programming Problem-II: Extension to a General Algorithm," Opns. Res. 16, 928-947.

Shapiro, J. F. (1968c), "Shortest Route Methods for Finite State Space Deterministic Dynamic Programming Problems," SIAM J. 16, 1232-1250.

Spielberg, K. (1969a), "Algorithms for the Simple Plant-Location Problem with Some Side Conditions," Opns. Res. 17, 85-111.

Spielberg, K. (1969b), "Plant Location with Generalized Search Origin," Man. Sci. 16, 165-178.

Srinivasan, A. V. (1965), "An Investigation of Some Computational Aspects of Integer Programming," JACM 12, 525-535.

Steinberg, D. I. (1970), "The Fixed Charge Problem," Nav. Res. Log. Quart. 17, 217-236.

Thangavelu, S. R. and C. M. Shetty (1971), "Assembly Line Balancing by 0-1 Integer Programming," AIIE Trans. III, 64-69.

Tomlin, J. A. (1970), "Branch and Bound Methods for Integer and Non-Convex Programming," 437-450, in Abadie (1970).

Tomlin, J. A. (1971), "An Improved Branch-and-Bound Method for Integer Programming," Opns. Res. 19, 1070-1074.

Toregas, C., R. Swain, C. Revelle, and L. Bergman (1971), "The Location of Emergency Service Facilities," Opns. Res. 19, 1363-1373.

Trauth, C. A. and R. E. Woolsey (1969), "Integer Linear Programming: A Study in Computational Efficiency," Man. Sci. 15, 481-493.

Trotter, L. E., Jr. and C. M. Shetty (1971), "An Algorithm for the Bounded Variable Integer Programming Problem," Dept. of Industrial Engineering, Georgia Institute of Technology.

Veinott, A. F., Jr. and G. B. Dantzig (1968), "Integral Extreme Points," SIAM Rev. 10, 371-372.

Weingartner, H. M. and D. N. Ness (1967), "Methods for the Solution of Multi-Dimensional 0/1 Knapsack Problems," Opns. Res. 15, 83-103.

Wolfe, P. and L. Cutler (1963), "Experiments in Linear Programming," 177-200 in Graves and Wolfe (1963).

Woolsey, R. E. (1972), "A Candle to St. Jude, or Four Real World Applications of Integer Programming," Interfaces 2, (Bulletin of TIMS) 20-27.

Young, R. D. (1965), "A Primal (All-Integer) Integer Programming Algorithm," J. Res. Nat. Bur. Stds. 69B, 213-250.

Young, R. D. (1968), "A Simplified Primal (All-Integer) Integer Programming Algorithm," Opns. Res. 16, 750-782.

R. S. Garfinkel
Graduate School of Management
University of Rochester
Rochester, New York 14627

G. L. Nemhauser
Department of Operations Research
College of Engineering
Cornell University
Ithaca, New York 14850

Supported, in part, by the National Science Foundation under Grant GK-3228X to Cornell University.

The Group Problems and Subadditive Functions

RALPH E. GOMORY AND ELLIS L. JOHNSON

1. Introduction

A. Inequalities based on the integer nature of some or all of the variables are useful in almost any algorithm for integer programming. They can furnish cut-offs for branch and bound or truncated enumeration methods, or cutting plane methods. In this paper we describe methods for producing such inequalities.

We will attempt to outline our general approach, taking the pure integer case first.

Consider a pure integer problem

(1) $$Ax = b\ , \quad x \geq 0$$

in which A is an $m \times (m+n)$ matrix, x is an integer $m+n$ vector, and b an m-vector. If we consider a basis B (in most applications this will be an optimal basis) we can write (1) as

$$Bx_B + Nx_N = bx_B \geq 0\ , \quad x_N \geq 0$$

where x_B is the m-vector of basic variables and x_N the non-basic n-vector. The usual transformed matrix [1, pages 75-80] corresponds to the equations

$$x_B + B^{-1}Nx_N = B^{-1}b\ ,\ x_B \geq 0\ ,\ x_N \geq 0\ , \qquad \text{or}$$

(2) $$y + N'z = b'\ ,\quad y \geq 0\ ,\ z \geq 0\ .$$

Taking the ith row we have

$$y_i + \sum_{j=1}^{j=n} n'_{i,j} z_m = b'_i\ .$$

We can form a new but related equation by reducing all coefficients modulo 1 and replacing the equality by equivalence modulo 1 . This yields

(3) $$\sum_{j=1}^{j=n} \mathfrak{F}(n'_{ij}) z_j \equiv \mathfrak{F}(b'_i) \qquad (\text{mod } 1)\ .$$

Now any integer vector (y, z) satisfying (2) automatically satisfies (3) , so that any inequality

$$\sum_{j=1}^{j=n} \pi_j z_j \geq \pi_0\ , \qquad \text{or} \qquad \pi \cdot z \geq \pi_0$$

which is satisfied by all solutions z to (3) is also satisfied by all solutions to (2) , i.e.

$$(0, \pi) \cdot (y, z) \geq \pi_0$$

holds for any integer vector (y, z) satisfying (2) .

The approach of this paper is to develop inequalities valid for all solutions to (2) by obtaining those valid for all solutions to the simpler equations like (3) .

More generally, we can proceed as follows, let ψ be a linear mapping sending the points of m-space into some other topological group S with addition. If we have an equation (like (2))

(4) $$\sum_j C_j x_j = C_0$$

in which the C_j and C_0 are m vectors, we can obtain a new equation by using the mapping ψ to obtain, by linearity,

(5) $$\sum_j \psi(C_j x_j) = \psi(C_0)$$

which is an equation involving a set of group elements in S, the elements $\psi(C_j x_j)$. For integer x_j, $\psi(C_j x_j) = \psi(C_j)x_j$, so equivalent group equations are

(6) $$\sum_j \psi(C_j)x_j = \psi(C_0) .$$

In the discussion leading up to equation (3) the C_j were the columns of the matrix (I, N') and ψ was the mapping that sends an m vector into the fractional part of its ith coordinate. The group S was the unit interval with addition modulo 1. Equation (3) was the equation (6).

Again, if $\pi \cdot x \geq \pi_0$ holds for all integer x satisfying (5) or (6) it holds for integer x satisfying (4) .

In this paper we study equations such as (6) and develop inequalities for their solutions which are then satisfied by the solutions to (4). Specifically we study the case where S is I , the unit interval mod 1, and develop inequalities for the equations:

(7) $$\sum_{u \in U} ut(u) = u_0$$

where U represents the set $\psi(C_j) \in I$ and $t(u)$ is a non-negative integer. Equation (7), which we refer to as the problem (or equation), $P(U, u_0)$, is merely (6) rewritten in a different notation.

Returning to equation (4) when some of the x_j are not restricted to be integer, a linear mapping ψ still gives another equation (5) satisfied by all solutions to (4). Thus, any solution to (4) satisfies the equation

$$\sum_j \psi(C_j x_j) = \psi(c_0) .$$

Just as before, if any x_j is required to be integer, then $\psi(C_j x_j) = \psi(C_j)x_j$. Let J_1 denote the subset of j for which x_j is required to be integer and J_2 be the j for which x_j is only required to be non-negative. Then, any solution to (4) with x_j integer for $j \in J_1$ satisfies

$$\sum_{j \in J_1} \psi(C_j)x_j + \sum_{j \in J_2} \psi(C_j x_j) = \psi(C_0) . \tag{8}$$

When ψ is the same (fractional) map used to derive (3), we rewrite (8) as

$$\sum_{j \in J_1} \mathfrak{F}(n'_{ij})z_j + \sum_{j \in J_2} \mathfrak{F}(n'_{ij}z_j) \equiv \mathfrak{F}(b'_i) \qquad (\text{mod } 1). \tag{9}$$

Consider $n_{ij}z_j$ for $j \in J_2$. If $n'_{ij} = 0$, then z_j does not really enter into the equation. If $n'_{ij} \neq 0$ we can rescale z_j by letting

$$z'_j = |n'_{ij}|z_j , \quad j \in J_2 .$$

Let $J_2^+ = \{j \in J_2 : n'_{ij} > 0\}$ and $J_2^- = \{j \in J_2 : n'_{ij} < 0\}$. Then $z'_j = n'_{ij}z_j$ for $j \in J_2^+$ and $-z'_j = n'_{ij}z_j$ for $j \in J_2^-$. The restriction $z_j \geq 0$ is equivalent to $z'_j \geq 0$. Hence, (9) becomes

$$\sum_{j \in J_1} \mathfrak{F}(n'_{ij})z_j + \sum_{j \in J_2^+} \mathfrak{F}(z'_j) - \sum_{j \in J_2^-} \mathfrak{F}(z'_j) \equiv \mathfrak{F}(b'_1) \quad (\text{mod } 1). \tag{10}$$

Since

$$\sum_{j \in J_2^+} \mathfrak{F}(z'_j) \equiv \mathfrak{F}\Big(\sum_{j \in J_2^+} z'_j\Big) \qquad (\text{mod } 1)$$

(10) can be simplified to

$$\sum_{j \in J_1} \mathfrak{F}(n'_{ij})z_j + \mathfrak{F}(z^+) - \mathfrak{F}(z^-) \equiv \mathfrak{F}(b'_i) \quad (\text{mod } 1) \tag{11}$$

where

$$z^+ = \sum_{j \in J_2^+} z'_j ,$$

$$z^- = \sum_{j \in J_2^-} z'_j .$$

We can rewrite (11) in a form similar to (7) to obtain the problem we call $P^+_-(U, u_0)$:

$$\sum_{u \in U} ut(u) + \mathfrak{F}(s^+) - \mathfrak{F}(s^-) = u_0 . \tag{12}$$

In this paper, we concentrate on the development of valid inequalities for equations of the form (7) and (12). These inequalities, satisfied by every solution to (7) or (12), are immediately appliciable to the original problem (4). In the case of an inequality

$$\sum_{j \in J_1} \pi_j z_j + \pi^+ z^+ + \pi^- z^- \geq 1 \tag{13}$$

satisfied by every solution to (11), the inequality

$$\sum_{j \in J_1} \pi_j z_j + \sum_{j \in J_2^+} (\pi^+ n'_{ij}) z_j + \sum_{j \in J_2^-} (-\pi^- n'_{ij}) z_j > 1 \tag{14}$$

is satisfied by every solution to (10), and hence to (4) .

2. Problem Definition

Let I be the group formed by the real numbers on the interval $[0,1)$ with addition modulo 1 . Let U be a subset of I and let t be an integer-valued function on U such that (i) $t(u) \geq 0$ for all $u \in U$, and (ii) t has a finite support; that is, $t(u) > 0$ only for a finite subset U_t of U .

The notation and definitions above will be used throughout so that t will always refer to a non-negative integer valued function with finite support.

We say that the function t is a solution to the problem $P(U, u_0)$, for $u_0 \in I - \{0\}$, if

$$\sum_{u \in U} ut(u) = u_0 . \tag{1}$$

Here, of course, addition and multiplication are taken modulo 1. Let $T(U, u_0)$ denote the set of all such solutions t to $P(U, u_0)$.

Correspondingly, <u>the problem</u> $P^{+}_{-}(U, u_0)$ <u>has solution</u> $t' = (t, s^+, s^-)$ satisfying

$$\sum_{u \in U} ut(u) + \mathfrak{F}(s^+) - \mathfrak{F}(s^-) = u_0 \tag{2}$$

where t is, as before, a non-negative integer valued function on U with a finite support, where s^+, s^- are non-negative real numbers, and where $\mathfrak{F}(x)$ denotes the element of I given by taking the fractional part of a real number x. Let $T^{+}_{-}(U, u_0)$ denote the set of solutions $t' = (t, s^+, s^-)$ to $P^{+}_{-}(U, u_0)$.

It is also possible to define problems $P^{+}(U, u_0)$ and $P_{-}(U, u_0)$ in which only s^+ or s^- appear, and these problems are useful in some situations. Their development parallels that of $P^{+}_{-}(U, u_0)$.

The notation $u \in I$ will mean that u is a member of the group I so that arithmetic is always modulo 1. If we want to consider u as a point on the real line with real arithmetic, we will write $|u|$. Thus, $|u|$ and $\mathfrak{F}(x)$ are mappings in opposite directions between I and the reals. And, in fact, $\mathfrak{F}(|u|) = u$ but x and $|\mathfrak{F}(x)|$ may differ by an integer.

<u>Definition 1.</u> <u>Valid Inequalities.</u> For any problem $P(U, u_0)$, we have so far defined the solution set $T(U, u_0)$. A <u>valid inequality for the problem</u> $P(U, u_0)$ is a real-valued function π defined for all $u \in I$ such that

$$\pi(u) \geq 0 \ , \quad \text{all } u \in I \ , \quad \text{and} \quad \pi(0) = 0 \ , \tag{3}$$

and

$$\sum_{u \in U} \pi(u)t(u) \geq 1 \ , \quad \text{all } t \in T(U, u_0) \ . \tag{4}$$

For the problem $P^{+}_{-}(U, u_0)$, $\pi' = (\pi, \pi^+, \pi^-)$ is a <u>valid inequality for</u> $P^{+}_{-}(U, u_0)$ when π is a real-valued function on I satisfying (3), and π^+, π^- are non-negative real numbers such that

$$(5) \qquad \sum_{u \in U} \pi(u)t(u) + \pi^+ s^+ + \pi^- s^- \geq 1 \text{, all } t' \in T^+_-(U, u_0) .$$

A valid inequality (π, π^+, π^-) for $P^+_-(I, u_0)$ can be used to give a valid inequality for $P(U, u_0)$ or $P^+_-(U, u_0)$ for any subset U of I. For example, $\Sigma\pi(u)t(u) \geq 1$ is clearly true for any $t \in T(U, u_0)$ since that t can be extended to a function t' belonging to $T(I, u_0)$ by letting $t'(u) = 0$ for $u \in$ i-U. Thus, the problem $P^+_-(I, u_0)$ acts as a master problem as in [2] where the master problem is a group problem with all group elements present. This fact is the main reason for studying the case $U = I$ in such detail. However, the next two properties of valid inequalities do not necessarily carry over to subsets U and I.

Definition 2. Minimal Valid Inequalities. A valid inequality π for $P(U, u_0)$ is a minimal valid inequality for $P(U, u_0)$ if there is no other valid inequality ρ for $P(U, u_0)$ satisfying $\rho(U) < \pi(U)$, where $\rho(U) < \pi(U)$ is defined to mean $\rho(u) \leq \pi(u)$ for all $u \in U$ and $\rho(u) < \pi(u)$ for at least one $u \in U$. A valid inequality π' for $P^+_-(U, u_0)$ is a minimal valid inequality for $P^+_-(U, u_0)$ satisfying $\rho'(u) < \pi'(u)$ where $\rho'(u) < \pi'(U)$ is defined to mean

$$\rho^+ \leq \pi^+ , \quad \rho^- \leq \pi^- ,$$

and

$$\rho(u) \leq \pi(u) , \quad u \in U ,$$

with strict inequality holding for at least one of the above inequalities.

The minimal valid inequalities are important because a valid inequality which is not minimal is implied by some other valid inequality. Notice that we have scaled the inequalities to have a right-hand side equal to one, and minimality is always with respect to that scaling.

Definition 3. Extreme Valid Inequalities. A valid inequality π for $P(U, u_0)$ is an extreme valid inequality for $P(U, u_0)$ if π can not be written as $\pi = \frac{1}{2}\rho + \frac{1}{2}\sigma$ for $\rho \neq \sigma$ where ρ, σ are valid inequalities for $P(U, u_0)$.

A valid inequality $\pi' = (\pi, \pi^+, \pi^-)$ for $P^{\pm}(U, u_0)$ is an extreme valid inequality for $P^{\pm}(U, u_0)$ if π' cannot be written as $\pi' = \frac{1}{2}\rho + \frac{1}{2}\sigma'$ for $\rho' \neq \sigma'$ where ρ', σ' are valid inequalities for $P^{\pm}(U, u_0)$.

Theorem I.1 of [3] says that the extreme valid inequalities are also minimal. These inequalities are in some sense "the best" possible since they cannot be derived from any other valid inequalities.

Definition 4. Subadditive Valid Inequalities. A valid inequality π for $P(U, u_0)$ is a subadditive valid inequality for $P(U, u_0)$ if

(6) $$\pi(u) + \pi(v) \geq \pi(u+v)$$

whenever all three of u, v, and $u+v$ are in U.

For a valid inequality π' for $P^{+}_{-}(U, u_0)$ to be subadditive, we require, in addition to (6),

(7) $$\pi(u) + \pi^+ |v-u| \geq \pi(v), \text{ whenever } u, v, \in U, |u| < |v|,$$

(8) $$\pi(u) + \pi^- |u-v| \geq \pi(v), \text{ whenever } u, v, \in U, |u| > |v| .$$

Theorems I.1 and I.2 of [3] prove the following sequence of inclusions: the set of valid inequalities include the subadditive valid inequalities which include minimal valid inequalities which include extreme valid inequalities. The subadditive valid inequalities form a convex set contained in the larger convex set of valid inequalities. Theorem I.3 of [3] says that the extreme points of the set of subadditive valid inequalities include all the extreme valid inequalities. Further, among the extreme subadditive valid inequalities, those which are extreme valid inequalities are the minimal ones. This fact allows us to actually find the extreme valid inequalities for some problems.

3. Subadditivity for Subgroups U .

The problems for which we can find extreme valid inequalities are $P(U, u_0)$ or $P_-^+(U, u_0)$ where U is a non-empty subgroup of I. We permit $U = I$ and note that 0 is always in U.

Definition 5. A function π defined on I is subadditive on a subgroup U of I if

$$\pi(u) \geq 0 \;, \quad u \in I \;, \quad \pi(0) = 0 \;, \quad \text{and}$$

$$\pi(u) + \pi(v) \geq \pi(u+v), u, v, \in U \,.$$

The function π is not assumed to be a valid inequality. Theorem I. 5 of [3] establishes the close connection between subadditive functions on U and subadditive valid inequalities. That theorem asserts that the subadditive valid inequalities for $P(U, u_0)$ are precisely the subadditive functions π satisfying $\pi(u_0) \geq 1$. Furthermore, if π is a subadditive function on U and $\pi(u_0) > 0$ for some $u_0 \in U$, then π^* defined by

$$\pi^*(u) = \frac{\pi(u)}{\pi(u_0)} \,, \quad u \in I \,, \tag{9}$$

is a valid inequality for $P(U, u_0)$.

The analogous theorem for $P_-^+(U, u_0)$ will now be developed.

Definition 6. $\pi' = (\pi, \pi^+, \pi^-)$ is an extended subadditive function on a subgroup U of I if π is subadditive on U, and if, in addition

$$\pi^+ |u| \geq \pi(u) \,, \quad u \in U \,, \tag{11}$$

$$\pi^- |u| \geq \pi(-u) \,, \quad -u \in U \,. \tag{12}$$

Theorem 1. 5B of [3] says that the subadditive valid inequalities are precisely the extended subadditive functions which satisfy both:

(13) $\pi(u) + \pi^{+}|u_0 - u| \geq 1$ whenever $u \in U$ and $|u| \leq |u_0|$,

(14) $\pi(u) + \pi^{-}|u - u_0| \geq 1$ whenever $u \in U$ and $|u| \leq |u_0|$,

Although subadditivity of π on I is not easily characterized, a graphical representation can be given. If π is drawn as a periodic function with period 1 as in figure 1, and if we then shift the image of π by transferring the origin to another point $(u, \pi(u))$, as shown in figure 1 by dotting lines, then subadditivity of π is equivalent to the dotted line staying above the solid line. The reason is that the points on the dotted line are of the form $(u + v, \pi(u) + \pi(v))$.

Some examples of subadditive functions are shown in figure 2. Figure 2(a) shows Gomory's fractional cut, figure 2(b) shows Gomory's mixed integer cut, and figure 2(c) is a more complicated function.

When π is subadditive on I, then (π, π^{+}, π^{-}) is an extended subadditive function on I if, and only if, π has a right derivative at 0 and a left derivative at 1, and π^{+} is larger than or equal to the right derivative at 0 while π^{-} is larger than or equal to the absolute value of the left derivative at 1. In figure 2(a), π has a right derivative at 0 equal to 1, but π has no finite left derivative at 1. In both figures 2(b) and (c), π has both left and right derivatives at 0 and 1, so π^{+}, π^{-} could be set to make (π, π^{+}, π^{-}) an extended subadditive function in those two cases.

4. Minimality for Subgroups U.

Theorem 1.6 of [3] is as follows: If U is a subgroup of I with $u_0 \in U$ and if π is a valid inequality for $P(U, u_0)$, then π is a minimal valid inequality if and only if

(15) $$\pi(u) + \pi(u_0 - u) = 1, \quad \text{all } u \in U.$$

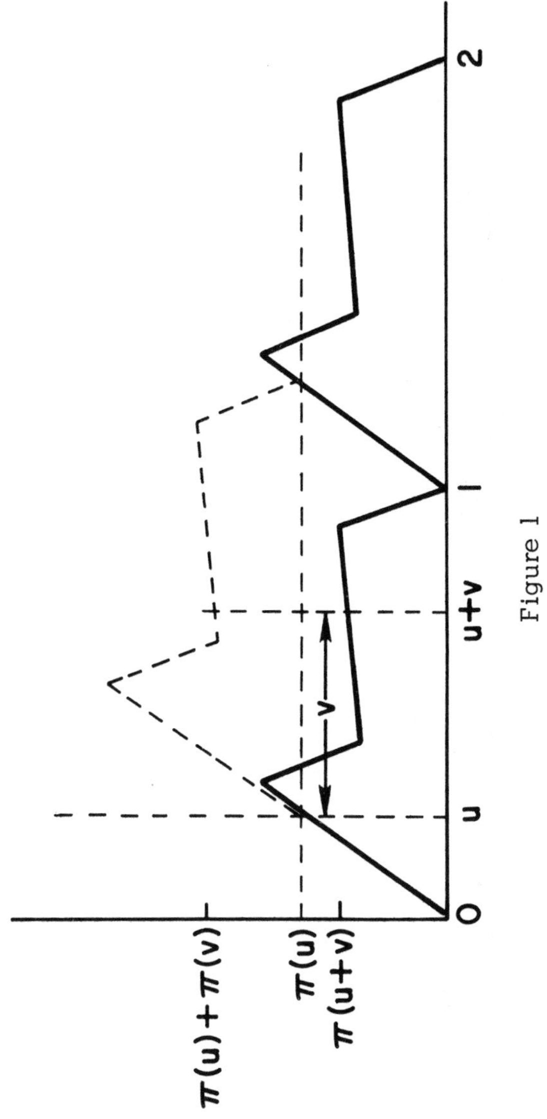

Figure 1

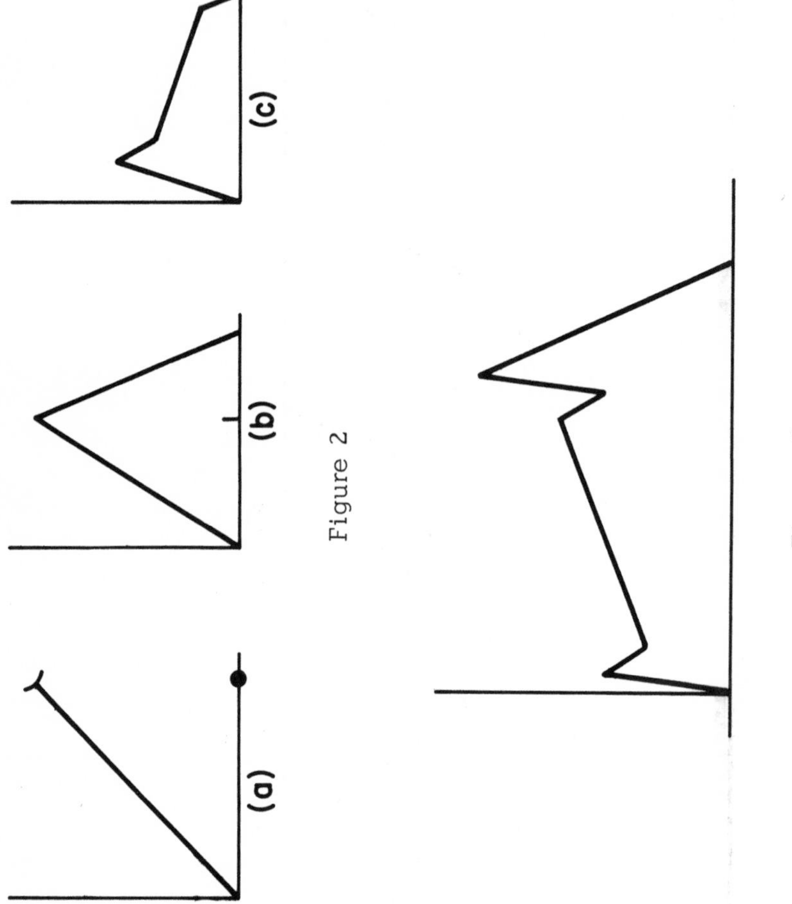

Figure 2

Figure 3

This condition imposes a peculiar symmetry on π so that $\pi(u)$ for $\frac{1}{2}u_0 < u < u_0$ is determined by $\pi(u)$ on $[0, \frac{1}{2}u_0]$, for example. This symmetry is illustrated in figure 3. One way of picturing it is that $\pi(u)$ and $\pi(u_0-u)$ must change by equal amounts but in opposite directions as u increases from 0 or as u decreases from 1.

5. $P(G_n, u_0)$, $u_0 \in G_n$

Let G_n denote the subset

$$G_n = \{0, \frac{1}{n}, \frac{2}{n}, \ldots, \frac{n-1}{n}\}$$

of I. The elements of G_n will be denoted $g_i = \mathfrak{F}(i/n)$. Each set G_n for $n \geq 1$ is a subgroup of I. By virtue of G_n being a subgroup, the results of 3 and 4 apply to this section.

The results from 3 and 4 are specialized in Theorem II. 2 of [3]: The extreme valid inequalities for $P(G_n, u_0), u_0 \in G_n$, are the extreme points of the solutions to

$$\pi(g_i) \geq 0, \quad \pi(0) = 0, \tag{16}$$

$$\pi(g_i) + \pi(g_j) \geq \pi(g_i + g_j), \tag{17}$$

$$\pi(u_0) \geq 1 \tag{18}$$

which satisfy the additional equations,

$$\pi(g_i) + \pi(u_0 - g_i) = 1, \quad g_i \in G_n.$$

In particular, (4) implies $\pi(u_0) = 1$ since $\pi(0) = 0$.

In [2], the extreme valid inequalities, or faces, are given for all G_n, $n = 1, \ldots, 11$. In addition, the faces are given for non-cyclic, but still abelian, groups of order less than 11. An example of the linear inequalities defining those faces is given below for $n = 6$ and $u_0 = g_5$:

$\pi(1)$	$\pi(2)$	$\pi(3)$	$\pi(4)$	$\pi(5)$	
1			1	-1	= 0
	1	1		-1	= 0
2	-1				≥ 0
1	1	-1			≥ 0
1		1		-1	≥ 0
	2			-1	≥ 0
-1		1		1	≥ 0
	-1			2	≥ 0

6. $P_-^+(G_n, u_0)$, $u_0 \in I$

The condition (2) now becomes

$$g_i t(g_1) + \ldots + g_{n-1} t(g_{n-1}) + \mathfrak{F}(s^+) - \mathfrak{F}(s^-) = u_0 ,$$

where $g_i = \mathfrak{F}(i/n)$ as before and where the $t(g_i)$ must be non-negative integers and s^+, s^- must be non-negative real values. We no longer confine u_0 to be in G_n. Let $L(u_0)$ and $R(u_0)$ denote, respectively, the points of G_n immediately below and above u_0. If u_0 happens to be in G_n, then $L(u_0) = R(u_0) = u_0$.

From Theorem II. 2B of [3] we know that the extreme valid inequalities π' for $P_-^+(G_n, u_0)$, $u_0 \in I$, are the extreme points of the solutions to the system of linear equations and inequalities (16), (17), and all of the following:

(19) $$\pi^+ \frac{1}{n} \geq \pi(g_1),\ g_1 = \mathfrak{F}(\frac{1}{n}) ,$$

(20) $$\pi^- \frac{1}{n} \geq \pi(g_n\text{-}1),\ g_n\text{-}1 = \mathfrak{F}((n-1)/n).$$

(21) $$\pi(L(u_0)) + \pi^+ |u_0 - L(u_0)| = 1 ,$$

(22) $$\pi(R(u_0)) + \pi^- |R(u_0) - u_0| = 1 ,$$

(23) for all $g_i \in G_n$, $\pi(g_i) + \pi(L(u_0)-g_i) = \pi(L(u_0))$

or $\pi(g_i) + \pi(R(u_0)-g_i) = \pi(R(u_0))$.

In [3], the extreme valid inequalities for $P^+_-(G_n, u_0)$ are given for $n = 1, \ldots, 7$ and all u_0. We give below the inequalities used to generate these faces for $n = 6$ and $u_0 = \frac{3}{4}$:

π^+	$\pi(1)$	$\pi(2)$	$\pi(3)$	$\pi(4)$	$\pi(5)$	π^-	
	2	-1					≥ 0
	1	1	-1				≥ 0
	1		1	-1			≥ 0
	1			1	-1		≥ 0
		2		-1			≥ 0
		1	1		1		≥ 0
	-1		1	1			≥ 0
		-1	1		1		≥ 0
		-1		2			≥ 0
			-1	1	1		≥ 0
$\frac{1}{6}$	-1						≥ 0
					-1	$\frac{1}{6}$	≥ 0
$\frac{1}{12}$				1			$= 1$
					1	$\frac{1}{12}$	$= 1$

In addition, condition (23) must be checked in order for (π, π^+, π^-) to represent an extreme valid inequality. In the appendix of [3], the computation and condition (23) are discussed.

7. Valid Inequalities for $P(U, u_0)$

We now connect the results about $P(G_n, u_0)$ with the general problem $P(U, u_0)$. Here, U can be any subset of the unit interval including the interval I itself. Theorem III.1 of [3] says that valid inequalities can be obtained simply by connecting the points $(g_n, \pi(g_n))$ by straight line segments. More precisely, if π is a subadditive function on G_n and if

$$(24) \qquad \pi(u) = n\{|u-L(u)|\pi(R(u)) + |R(u)-u|\pi(L(u))\}, \quad u \in I-G_n,$$

Then, π is a subadditive function on I, and π^* defined on I by

$$\pi^*(u) = \frac{\pi(u)}{\pi(u_0)}, \quad u \in I,$$

is a valid inequality for any $P(U, u_0)$, U a subset of I, provided $\pi(u_0) > 0$.

8. Valid Inequalities for $P_-^+(U, u_0)$

From valid inequalities for $P_-^+(G_n, u_0)$, a different method for generating valid inequalities for $P_-^+(U, u_0)$ is available. This method will be referred to as the <u>two-slope fill-in</u> (Theorem III.3 [3]). Let $\pi' = (\pi, \pi^+, \pi^-)$ be an extended subadditive function on G_n. Define $\pi(u)$ for $u \in I-G$ by

$$(25) \qquad \pi(u) = \min\{\pi(L(u)) + \pi^+|u-L(u)|, \\ \pi(R(u)) + \pi^-|R(u)-u|\}.$$

Then, π' is an extended subadditive function on I, and ρ' defined by

$$\rho' = \frac{1}{\pi(u_0)}(\pi, \pi^+, \pi^-)$$

is a valid inequality for $P_-^+(U, u_0)$ provided $\pi(u_0) > 0$.

Theorem II. 2B of [3] shows how to compute faces for $P_{-}^{+}(G_n, u_0)$ and this theorem shows how to use them to generate valid inequalities for any U. Table 2 of [3] was obtained using Theorem II. 2B, and we will frequently refer to the two-slope fill-in of those faces.

The functions π generated by this two-slope fill-in can also be used for $P(U, u_0)$ as well as $P_{-}^{+}(U, u_0)$. That is, for a problem with neither s^+ nor s^-, the π can be used, ignoring π^+ and π^-, to give a valid inequality for $P(U, u_0)$. The functions π generated by the two-slope fill-in have the advantage over the straight line fill-in that they are generated for a particular u_0 so that $\pi(u_0)$ will be large and the resulting inequality stronger.

<u>Example 1:</u> Consider the integer linear program

$$x_j \geq 0, \; x_j \text{ integer}, \quad j = 1,2,3,4,5$$

$$x_1 + 2x_2 + x_3 + x_4 + 5x_5 = 10$$

$$3x_1 - 3x_2 + 2x_3 - 3x_4 + 3x_5 = 5$$

$$x_1 + x_2 + x_3 + 2x_4 + 2x_5 = Z(\min).$$

The optimum linear programming tableau is

$$x_1 + \frac{7}{9}x_3 - \frac{1}{3}x_4 + 2\frac{1}{3}x_5 = 4\frac{4}{9}$$

$$x_2 + \frac{1}{9}x_3 + \frac{2}{3}x_4 + 1\frac{1}{3}x_5 = 2\frac{7}{9}$$

$$\frac{1}{9}x_3 + 1\frac{2}{3}x_4 + \frac{1}{3}x_5 = Z(\min).$$

The optimum linear programming solution is $x_1 = 4\frac{4}{9}$, $x_2 = 2\frac{7}{9}$, $x_3 = x_4 = x_5 = 0$, $z = 7\frac{2}{9}$. From the first row of the tableau, using as the mapping $\psi : \psi(A^j x_j) = \mathfrak{F}(a_{ij}) x_j$, we obtain

$$\frac{7}{9}x_3 + \frac{2}{3}x_4 + \frac{1}{3}x_5 = \frac{4}{9} (\bmod 1).$$

That is, $U = \{7/9, 2/3, 1/3\}$, $u_0 = 4/9$. The mapping used in the introduction is simply this; that is, we get a problem $P(U, u_0)$ from every row of an optimum linear programming tableau for which the basic variable is integer constrained but at a fractional value.

From Appendix 5 of [2], we find the following three extreme valid inequalities among those for $P(G_n, u_0)$, $n = 2, 3, 6$.

$$P(G_2, \tfrac{1}{2}): \ \pi_1(0) = 0\ , \quad \pi_1(\tfrac{1}{2}) = 1\ ;$$

$$P(G_3, \tfrac{1}{3}): \ \pi_2(0) = 0\ , \quad \pi_2(\tfrac{1}{3}) = 1\ , \quad \pi_2(\tfrac{2}{3}) = \tfrac{1}{2}\ ;$$

$$P(G_6, \tfrac{1}{2}): \ \pi_3(0) = 0\ , \quad \pi_3(\tfrac{1}{6}) = \tfrac{1}{3}\ , \quad \pi_3(\tfrac{2}{6}) = \tfrac{2}{3}\ ,$$

$$\pi_3(\tfrac{3}{6}) = 1\ , \quad \pi_3(\tfrac{4}{6}) = \tfrac{1}{3}\ , \quad \pi_3(\tfrac{5}{6}) = \tfrac{2}{3}\ .$$

We could take any of the faces for cyclic groups from Appendix 5 and use them in the following way. The linear interpolation of extends π_1, π_2, π_3 to the interval I:

$$\pi_1(u) = \begin{cases} 2u\ , & 0 \le u \le \frac{1}{2} \\ 2-2u, & \frac{1}{2} < u < 1 \end{cases}$$

$$\pi_2(u) = \begin{cases} 3u\ , & 0 \le u \le \frac{1}{3} \\ \frac{3}{2} - \frac{3}{2}u, & \frac{1}{3} < u < 1 \end{cases}$$

$$\pi_3(u) = \begin{cases} 2u, & 0 \le u \le \frac{1}{2} \\ 3-4u, & \frac{1}{2} < u < \frac{2}{3} \\ -1+2u, & \frac{2}{3} < u \le \frac{5}{6} \\ 4-4u, & \frac{5}{6} < u < 1 \end{cases}$$

Our congruence problem has $u_0 = 4/9$, and so $\pi_1(u_0) = 8/9$, $\pi_2(u_0) = 5/6$, $\pi_3(u_0) = 8/9$. Since $U = \{7/9, 2/3, 1/3\}$, the valid inequalities from π_1, π_2, and π_3 are of the form

$$\frac{\pi_i(\frac{7}{9})}{\pi_i(u_0)}x_3 + \frac{\pi_i(\frac{2}{3})}{\pi_i(u_0)}x_4 + \frac{\pi_i(\frac{1}{3})}{\pi_i(u_0)}x_5 \geq 1 ,$$

for $i = 1, 2, 3$, and are given below:

$$\frac{1}{2}x_3 + \frac{3}{4}x_4 + \frac{3}{4}x_5 \geq 1 ;$$

$$\frac{2}{5}x_3 + \frac{3}{5}x_4 + \frac{6}{5}x_5 \geq 1 ;$$

$$\frac{5}{8}x_3 + \frac{3}{8}x_4 + \frac{3}{4}x_5 \geq 1 .$$

The 'fractional cutting plane' is, here,

$$\frac{7}{9}x_3 + \frac{2}{3}x_4 + \frac{1}{3}x_5 \geq \frac{4}{9}, \text{ or } \frac{7}{4}x_3 + \frac{3}{2}x_4 + \frac{3}{4}x_5 \geq 1 .$$

That inequality is obtained from the subadditive function π on I given by $\pi(u) = u$. The figure 4 illustrates the functions π, π_1, π_2, π_3.

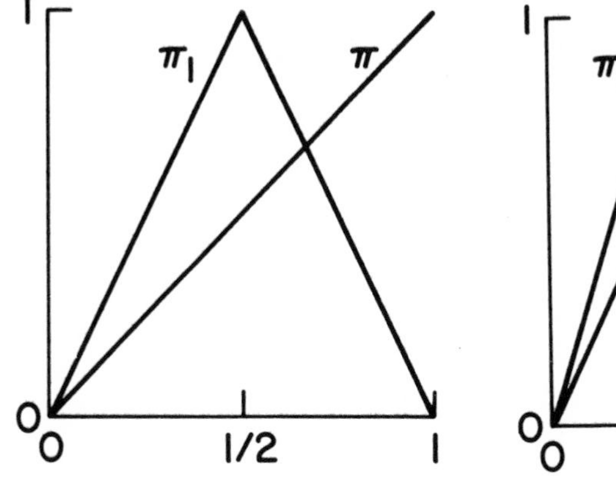

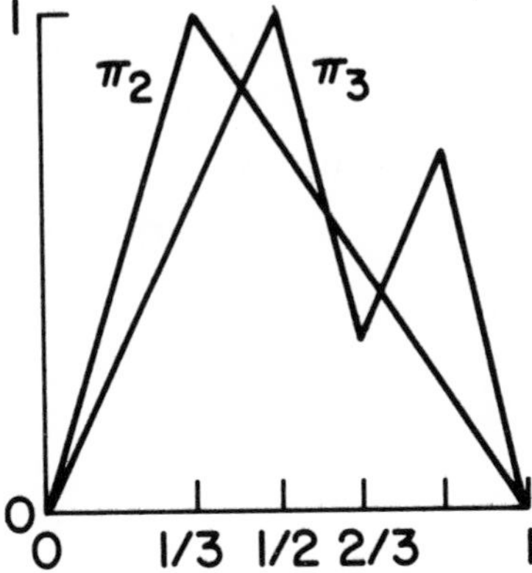

Figure 4

Example 2: Consider the same integer program but without the integrality restriction on x_5. The first row of the optimal linear programming tableau now gives the congruence:

$$\frac{7}{9}x_3 + \frac{2}{3}x_4 + s^+ \equiv \frac{4}{9} \pmod 1$$

where $s^+ = \frac{7}{3}x_5$. Thus, $U = \{\frac{7}{9}, \frac{2}{3}\}$ and $u_0 = \frac{4}{9}$.

From table 2 of the appendix of [3], for $n = 1$ the only extreme valid inequality has

$$\pi^+ = \frac{1}{|u_0|}, \quad \pi^- = \frac{1}{1 - |u_0|} .$$

Here, $u_0 = 4/9$ so $\pi^+ = 9/4$ and $\pi^- = 9/5$. Another extreme valid inequality, this time for $n = 3$, is

$$\pi(\tfrac{1}{3}) = \frac{1}{3|u_0|}, \quad \pi(\tfrac{2}{3}) = \frac{1}{6|u_0|}, \quad \pi^+ = \frac{1}{|u_0|}, \quad \pi^- = \frac{6|u_0|-1}{4|u_0|-6|u_0|^2} .$$

Since $u_0 = 4/9$ here,

$$\pi(\tfrac{1}{3}) = \frac{3}{4}, \quad \pi(\tfrac{2}{3}) = \frac{3}{8}, \quad \pi^+ = \frac{9}{4}, \quad \pi^- = \frac{45}{1.6} .$$

The two-slope fill-in extends these two inequalities to functions π_1 and π_2 on the unit interval:

$$\pi_1(u) = \begin{cases} \frac{9}{4}|u| & , \; 0 \le |u| \le \frac{4}{9} \\ \frac{9}{5}(1-|u|) , & \frac{4}{9} \le |u| \le 1 , \end{cases}$$

$$\pi_2(u) = \begin{cases} \frac{9}{4}|u|, & 0 \le |u| \le \frac{4}{9}, \\ \frac{3}{8} + \frac{45}{16}(\frac{2}{3} - |u|), & \frac{4}{9} \le |u| \le \frac{2}{3} \\ \frac{3}{8} + \frac{9}{4}(|u| - \frac{2}{3}), & \frac{2}{3} \le |u| \le \frac{7}{9} \\ \frac{45}{16}(1 - |u|), & \frac{7}{9} \le |u| \le 1 \end{cases}$$

Since $U = \{\frac{7}{9}, \frac{2}{3}\}$, a valid inequality is

$$\pi_i(\frac{7}{9})x_3 + \pi(\frac{2}{3})x_4 + \pi_i^+ s^+ \ge 1, \quad i = 1, 2, \qquad \text{or}$$

$$\pi_i(\frac{7}{9})x_3 + \pi_i(\frac{2}{3})x_4 + \frac{7}{3}\pi_i^+ x_5 \ge 1 \quad i = 1, 2 .$$

Evaluating π_i at $\frac{7}{9}$ and $\frac{2}{3}$ gives the two valid inequalities

$$\frac{2}{5}x_3 + \frac{3}{5}x_4 + \frac{21}{4}x_5 \ge 1, \quad \text{and}$$

$$\frac{5}{8}x_3 + \frac{3}{8}x_4 + \frac{21}{4}x_5 \ge 1 .$$

Other inequalities can be generated in the same way from Table 2 of [3] .

Example 3: Consider the integer program from Example 1, but let us use the functions π_1, π_2 from Example 2 to give cutting planes for that pure integer program. This example will provide a comparison on the coefficient of a variable (here x_5) which is an integer variable in one case and a continuous variable in another.

Now $U = \{\frac{7}{9}, \frac{2}{3}, \frac{1}{3}\}$ as in Example 1, and the valid inequalities from π_1 and π_2 given in Example 2 are:

$$\frac{2}{5}x_3 + \frac{3}{5}x_4 + \frac{3}{4}x_5 \ge 1, \qquad \text{and}$$

$$\frac{5}{8}x_3 + \frac{3}{8}x_4 + \frac{3}{4}x_5 \geq 1 .$$

Notice that the coefficients for x_5 are smaller here than in Example 2, illustrating the additional strength gained by the integrality assumption on x_5.

Figure 5 shows π_1 and π_2 used here.

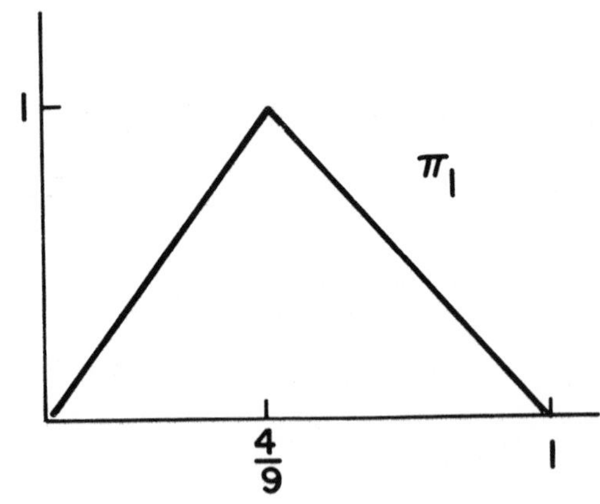

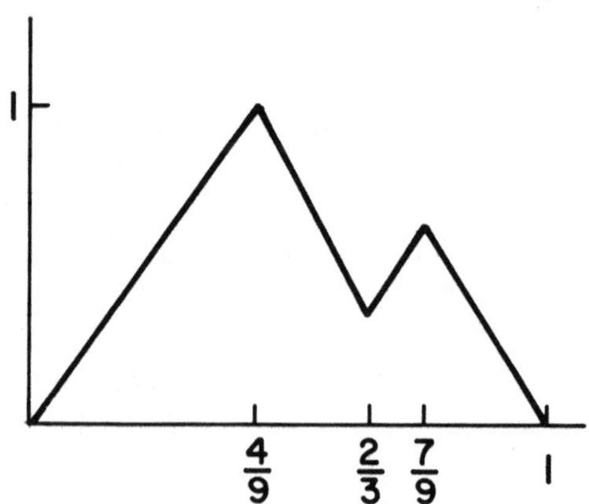

Figure 5

9. Rounding Methods

This section will give some ways to approximate the answer to a given cyclic group problem by a smaller group. For any problem:

$$x_j \geq 0 \text{ and integer}, \quad s^+ \geq 0, \; s^- \geq 0$$

$$\sum_{j=1}^{n} f_j x_j + s^+ - s^- \equiv f_0 \pmod 1$$

$$\sum_{j=1}^{n} c_j x_j + c^+ s^+ + c^- s^- = z ,$$

form the points (f_j, c_j) in the plane. The rounding methods are based on forming a function below the points and with no slope greater than c^+ nor less than $-c^-$. This function is then lowered to one which is subadditive by solving a smaller cyclic group problem. The derived function does give a valid inequality for the original group problem, and the value of the function at f_0 is a lower bound on the value of the cost z to satisfy the group problem.

When s^+ (or s^-) is not present in the problem, the restriction that no slope be greater than c^+ (or less than c^-) can be dropped.

The simplest method is first shown when neither s^+ nor s^- is present. In that case, there are no restrictions on the slopes. The method is described below. The number H can be any positive integer. Define

$$\gamma_h = \min_j \{c_j : \frac{h-1}{H} < f_j < \frac{h+1}{H}\}$$

for $h = 1, 2, \ldots, H-1$. Let $\gamma_0 = \gamma_H = 0$. Now, solve the cyclic group problem

$$\text{minimize} \sum_{h=1}^{H-1} \gamma_h t_h$$

$$\sum_{h=1}^{H-1} h t_h \equiv h_0 \pmod{H}$$

$$t_h \geq 0 \text{ and integer,}$$

where h_0 is either the integer above or below Hf_0.

Shortest path methods, including the one giving in Section 5, will give numbers $d(h) \leq \gamma_h$ such that

$$d(h) + d(g) \geq d(h + g \pmod{H}) .$$

Extend d to a function ρ on the unit interval by letting

$$\rho(x) = \max \{d(\lceil Hx \rceil), d(\lfloor Hx \rfloor)\}$$

where $\lceil y \rceil$ means the smallest integer larger than or equal to y and $\lfloor y \rfloor$ means the largest integer less than or equal to y. Figure 5 illustrates this function. The claim is that

$$\sum_{j=1}^{n} \rho(f_j) x_j \geq \rho(f_0)$$

is a valid inequality for the original problem. Furthermore, $\rho(f_0)$ is a lower bound on z for the original problem, and $\rho(f_0) \geq d(h_0)$.

The original choice of γ_h assures that $\rho(f_j) \leq c_j$. Thus, if the inequality is valid, the bound of $\rho(f_0)$ is true. All that remains is to show that ρ is subadditive, that is, $\rho(x) + \rho(y) \geq \rho(x+y \pmod{1})$. But,

$$\rho(x) + \rho(y) = \max \{d(\lceil Hx \rceil), d(\lfloor Hx \rfloor)\}$$

$$+ \max \{d(\lceil Hy \rceil, d(\lfloor Hy \rfloor)\} ,$$

and $\rho(x+y(\text{mod}))$ is either

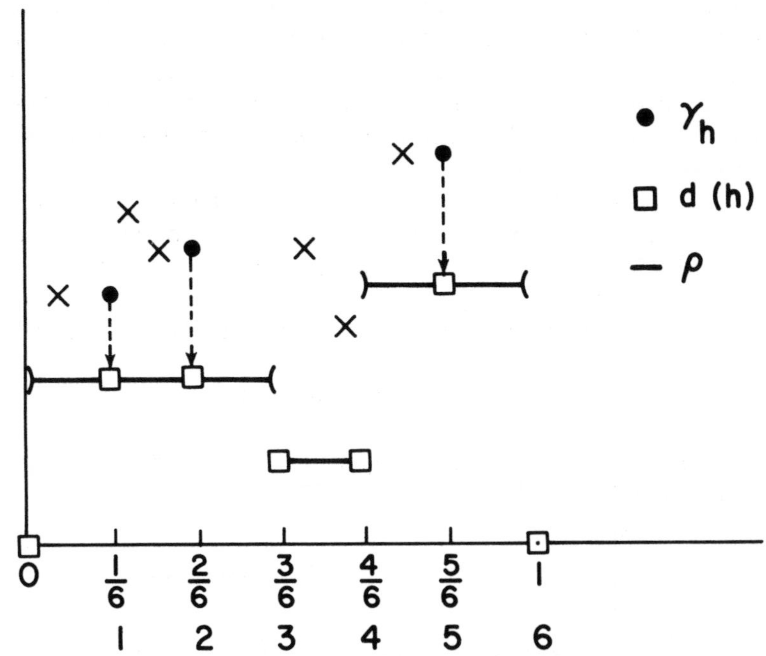

Figure 6

$$\max\{d(\lceil Hx\rceil + \lceil Hy\rceil),\ d(\lceil Hx\rceil + \lfloor Hy\rfloor)\}$$

or

$$\max\{d(\lceil Hx\rceil + \lfloor Hy\rfloor),\ d(\lfloor Hx\rfloor + \lfloor Hy\rfloor)\}\ ,$$

where the $+$ here is mod H. Which of the two terms is $\rho(x+y(\text{mod } 1))$ depends on where $Hx+Hy(\text{mod } H)$ happens to fall relative to $\lfloor Hx\rfloor + \lfloor Hy\rfloor$, $\lceil Hx\rceil + \lfloor Hy\rfloor$, $\lceil Hx\rceil + \lceil Hy\rceil$. However, any one of the numbers $d(h_1+h_2)$, for $h_1 = \lfloor Hx\rfloor$ or $\lceil Hx\rceil$ and $h_2 = \lceil H_y\rceil$ or $\lceil H_y\rceil$, is less than or equal to $\rho(x) + \rho(y)$ because that sum is the sum of two maxima including $d(h_1)$ and $d(h_2)$.

This method should help to establish the general principle. The general method involves extending the numbers $d(h)$, $h = 0,\ldots,H$, to a function $\rho(x)$, $0 \le x \le 1$, such that $\rho(h/H) = d(h)$. Further, the function ρ should be subadditive on the whole interval $[0,1]$ whenever d is subadditive on $0,1,\ldots,h$. Three general forms of such extensions are given below.

Given values $0 = d(0), d(1),\ldots,d(H-1), D(H) = 0$, and $d^+ > 0$, $d^- > 0$, such that $d(1) \le d^+/H$ and $d(H-1) \le d^-/H$, let $L(x) = \lfloor Hx \rfloor$ and $R(x) = \lceil Hx \rceil$ for $0 \le x \le 1$. Define

$$\rho_1(x) = \min\{d(L(x)) + d^+(x-L(x)),\ d(R(x)) + d^-(R(x)-x),$$

$$\max\{(L(x)),\ d(R(x))\}\} .$$

Here, the d are assumed subadditive, i.e., $d(h) + d(g) \ge d(g+h \pmod H)$.

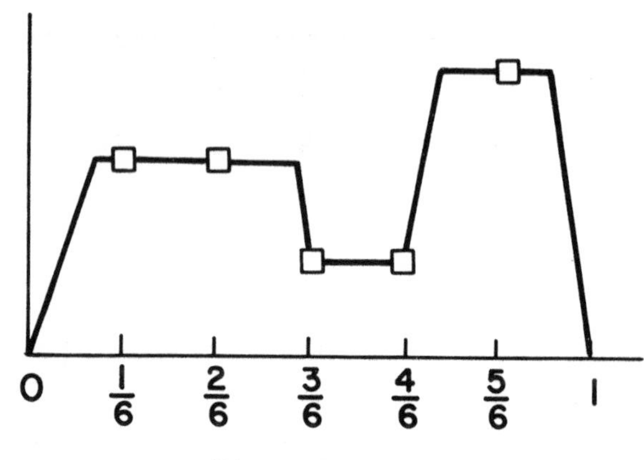

Figure 7

In order for ρ_1 to be continuous, $d(h-1) \le d(h) + d^-/H$ and $d(h+1) \le d(h) + d^+/H$ are required. These inequalities follow from subadditivity of d and from $d(1) \le d^+/H$, $d(H-1) \le d^-/H$ since then $d(h-1) \le d(h) + d(H-1) \le d(h) + d^-/H$. Then ρ_1 will be subadditive with slope from above at most d^+ and slope from below at least $-d^-$. To show ρ_1 subadditive, if $\rho_1(x)$ and $\rho_1(y)$ are given by $\max\{d(L(x)), d(R(x))\}$ and

$\max\{d(L(y)), d(R(y))\}$ respectively, then the previous proof still holds. Suppose either $\rho_1(x)$ or $\rho_1(y)$ is given by a term such as $d(L(x)) + d^+(x-L(x))$. In this case, x can be decreased to $L(x)$, decreasing $x+y$. But $\rho_1(x)$ decreases by more than or the same as $\rho_1(x+y)$ because nowhere is the slope steeper than d^+. Thus, it suffices to show that

$$d(L(x)) + \rho_1(y) \geq \rho_1(L(x) + y) .$$

If $\rho_1(y)$ is given by $d(L(y))+ d^+(y-L(y))$ or $d(R(y))+ d^-(R(y)-y)$, then y can be similarly moved to either $L(y)$ or $R(y)$, and subadditivity of ρ_1 follows from subadditivity of d. If $\rho_1(y)$ is given by $\max\{d(L(y)), d(R(y))\}$, then the result follows from subadditivity of d and from

$$\rho_1(L(x)+y) \leq \max\{d(L(x) + d(L(y)) ,$$

$$d(L(x)) + d(R(y))\} .$$

The simplified method given when no s^+ nor s^- is present can be easily modified by letting $d^+ = c^+$, $d^- = c^-$ and replacing γ_1 by $\min\{\gamma_1, d^+/H\}$ and γ_{H-1} by $\min\{\gamma_{H-1}, d^-/H\}$. Once this is done, there is some possibility of increasing some of the γ_h's. The idea is to first extend $\gamma_h, h = 0,\ldots,H$ to a function ρ_1 of this form. As long as $\rho_1(f_j) \leq c_j$, the resulting $\rho(f_0)$ will be a bound. Thus, the γ_h can be increased as long as the resulting ρ_1 satisfied $\rho_1(f_j) \leq c_j$. Then, reduction of γ_h to subadditive d_h results in a reduction of ρ_1 to a subadditive function and, thus, a valid inequality.

There are two other methods of extending subadditive d_h, $h = 0,\ldots,H$, to a subadditive function. These extensions were discussed in Section 6 but will be restated here. Both extensions give rise to rounding procedures along the lines given here. One extension is simply linear interpolation:

$$\rho_2(x) = \lambda d(L(x)) + (1-\lambda)d(R(x))$$

where $\lambda = (R(x)-x)/H$. Here, $d(1) \leq c^+/H$ and $d(H-1) \leq c^-/H$ are also needed when s^+ and s^- are present.

The other extension is a two-slope extension:

$$\rho_3(x) = \min\{d(L(x)) + d^+(x-L(x))),\\ d(R(x)) + d^-(R(x)-x)\}\ .$$

When s^+ and s^- are present, $d^+ \leq c^+$ and $d^- \leq c^-$ are required. Otherwise, d^+ and d^- are arbitrary positive numbers.

REFERENCES

1. Dantzig, G. B., Linear Programming and Extensions, Princeton University Press, Princeton, New Jersey, 1963.

2. Gomory, R. E., "Some Polyhedra Related to Combinatorial Problems", Linear Algebra and its Applications, Vol. 2(1969), pp. 451-558.

3. Gomory, R. E., and E. L. Johnson, "Some Continuous Functions Related to Corner Polyhedra I", Mathematical Programming, Vol. 3 (1972), pp. 23-86.

Mathematical Sciences Department
IBM Watson Research Center
Yorktown Heights, New York 10598

Cyclic Groups, Cutting Planes, Shortest Paths

ELLIS L. JOHNSON

1. Introduction

An integer program is the problem of minimizing z subject to:

$$x_j \geq 0 \ , \quad j = 1, \ldots, n+m \ ; \tag{1}$$

$$\sum_{j=1}^{n+m} a_{ij} x_j = b_i \ , \quad i = 1, \ldots, m \ ; \tag{2}$$

$$x_j \quad \text{integer for} \quad j \in J_I \ ; \tag{3}$$

$$z = \sum_{j=1}^{n+m} c_j x_j \ . \tag{4}$$

If the problem is solved as a linear program, some variables x_k for $k \in J_I$ will typically be basic and at value x_k^0 which are not integer. The canonical form (page 75, [1]) with respect to that optimum basis provides an equation

$$x_k + \sum_{j \in J_N} \bar{a}_{ij} \, x_j = x_k^0 \tag{5}$$

expressing x_k as a linear combination of the redundant variables x_j , $j \in J_N$, where J_N is the index set of the non-basic variables.

Relaxing the restriction $x_k \geq 0$ but retaining the integrality restriction on x_k in (5) gives a cyclic group problem:

$$\sum_{j \in J_N} \bar{a}_{ij} x_j \equiv x_k^0 \pmod{1}. \tag{6}$$

The restriction (6) says that x_k^0 and $\Sigma\, \bar{a}_{ij} x_j$ should differ by an integer. Then x_k given by (5) will be integer.

If every x_j, $j \in J_N$, should be integer-valued; that is, if $J_N \subset J_I$, then (6) can be replaced [3] by

$$\sum_{j \in J_N} f_j x_j \equiv f_0 \pmod{1} \tag{7}$$

when f_j is the fractional part of $\bar{a}_{ij}$ and f_0 is the fractional part of x_k^0. By the <u>fractional part</u> of the x_k^0 is meant a number f, $0 \leq f < 1$, such that $f + h = x_k^0$ for some integer h. Since $\bar{a}_{ij}$, and hence f_j and f_0, are rational, we can multiply (7) by a large enough integer γ to give

$$\sum_{j \in J_N} g_j x_j \equiv g_0 \pmod{\gamma}, \tag{8}$$

where g_j and g_0 are integers. The polyhedra of the convex hull of non-negative integer solutions to (8), and hence to (7), are the polyhedra corresponding to cyclic groups in [2].

When some x_j, $j \in J_N$, are not required to be integer-valued, (6) can be written as

$$\sum_{j \in J_1} f_j x_j + \sum_{j \in J_2} \bar{a}_{ij} x_j \equiv f_0 \pmod{1} \tag{9}$$

where $J_1 = J_N \cap J_I$ and $J_2 = J_N - J_I$ and, as in (7), f_j is the fractional part of $\bar{a}_{ij}$ and f_0 is the fractional part of x_j^0. Let

$$T = \{(x_j, j \in J_N): (9) \text{ holds for } x\}$$

and let $C(T)$ be the convex closure of T .

To define $C(T)$ as the solution set of a system of linear inequalities, let us define a class of functions $\Pi(f_0)$ on the unit interval. Here f_0 is any real number satisfying $0 < f_0 < 1$. A function π defined for real u such that $0 \leq u \leq 1$ is in $\Pi(f_0)$ if the following four conditions hold:

(10) $\pi(u) \geq 0 \ , \ 0 \leq u \leq 1 \ ,$ and $\pi(0) = \pi(1) = 0$;

(11) $\pi(u) + \pi(v) \geq \pi(u+v), \quad 0 \leq u, v \leq 1,$ where $u + v$ is taken modulo 1;

(12) $\pi(u) + \pi(f_0 - u) = \pi(f_0),$ for $0 \leq u \leq 1$, where $f_0 - u$ is taken modulo 1; and

(13) $$\pi^+ = \lim_{u \downarrow 0} \frac{\pi(u)}{u} \quad \text{and} \quad \pi^- = \lim_{u \uparrow 1} \frac{\pi(u)}{1-u}$$

both exist and are finite.

Condition (11) is referred to as <u>subadditivity</u> of π .

The following theorem (Theorem 1.5 and Property IV.7 of [3]) describes $C(T)$ as an intersection of half-planes. Let $n = |J_N|$.

<u>Theorem 1:</u>

$$C(T) = R_n^+ \cap \Big[\bigcap_{\pi \in \Pi(f_0)} \{(x_j, j \in J_N) : \pi(f_0) \leq \sum_{j \in J_1} \pi(f_j) x_j + \sum_{j \in J_2^+} \pi^+ \bar{a}_{ij} x_j - \sum_{j \in J_2^-} \pi^- \bar{a}_{ij} x_j \} \Big]$$

where $J_2^+ = J_2 \cap \{j : \bar{a}_{ij} > 0\}$ and $J_2^- = J_2 \cap \{j : \bar{a}_{ij} < 0\}$. If J_2^+ is empty, $\pi^+ < \infty$ should be dropped as a restriction on $\pi \in \Pi(f_0)$, and if J_2^- is empty, $\pi^- < \infty$ should be dropped. Condition (12) is not needed for the inequality resulting from π in the theorem to be valid. Instead, (12) is a property of

π which is imposed in order to give a better inequality.

Although the class of all $\pi \in \Pi(u_0)$ is obviously very large, for a given T only a finite number are needed to describe $C(T)$. Regardless of the number, Theorem 1 gives an optimality criterion for $x \in T$. For our present purpose, that criterion is all that is needed.

Theorem 2: For any objective function $z = \Sigma\, c_j x_j$ with $c_j \geq 0$ for $j \in J_N$, $x^* \in T$ minimizes z over all $x \in T$ provided

$$\sum_{j \in J_N} c_j x_j^* = \pi(f_0) \tag{14}$$

for some $\pi \in \Pi(f_0)$ satisfying

$$\pi(f_j) \leq c_j\,, \quad j \in J_1\,; \tag{15}$$

$$\pi^+ \bar{a}_{ij} \leq c_j\,, \quad j \in J_2^+\,; \tag{16}$$

$$-\pi^- \bar{a}_{ij} \leq c_j\,, \quad j \in J_2^-\,. \tag{17}$$

Proof: Since any $x \in T$ satisfies the inequality

$$\pi(f_0) \leq \sum_{j \in J_1} \pi(f_j) x_j + \sum_{j \in J_2^+} \pi^+ \bar{a}_{ij} x_j - \sum_{j \in J_2^-} \pi^- \bar{a}_{ij} x_j \tag{18}$$

used in Theorem 1, obviously $C(T) \subseteq \{(x_j, j \in J_N) : x_j \geq 0$ and (18) holds$\}$. Hence the minimum of z over this latter set is less than or equal to the minimum over $C(T)$, which in turn is equal to the minimum over T. The minimum of z subject to the inequality (18) with non-negative coefficients is equal to $\pi(f_0)$ times the minimum ratio c_j divided by the constraint coefficient in (18). By (15), (16), and (17) this ratio is greater than or equal to one, so the minimum of z subject to (18) is greater than or equal to $\pi(f_0)$. Putting these relations together gives

$$\min \Big\{ \sum_{j \in J_N} c_j x_j : x \in T \Big\} \geq \{\pi(f_0) : \pi \in \Pi(f_0) \text{ and } \pi \text{ satisfies (15), (16) and (17)}\}\,.$$

Clearly if the equality (14) can be achieved by some $x^* \in T$ and some such π, then optimality will be assured.

2. Algorithm

The algorithm will produce a function $\pi \in \Pi(f_0)$ satisfying (15), (16), and (17) and a solution $x^* \in T$ such that (14) holds. The function π will be piecewise linear with breakpoints $0 = e_0 < e_1 < \ldots < e_L = 1$ and $\pi(0) = \pi(1) = 0$. Each breakpoint is either increasing or fixed. Initially π has breakpoints 0, f_0, 1 and $\pi(0) = \pi(f_0) = \pi(1) = 0$. The breakpoints 0 and 1 are fixed, and f_0 is increasing. The breakpoints will always be in pairs: one e_i fixed and $e_j = f_0 - e_i$ (modulo 1) with e_j an increasing breakpoint.

Each iteration of the algorithm begins by increasing $\pi(e)$ for every increasing breakpoint e to $\pi(e) + \delta$, where δ is the same for each increasing breakpoint. For fixed breakpoints e, $\pi(e)$ remains fixed in value, and for u, $0 < u < 1$, $\pi(u)$ is determined by linear interpolation from the two neighboring breakpoints e_i, e_{i+1} such that $e_i \leq u < e_{i+1}$.

Increasing π will be called the LIFT step of the algorithm. What first prevents us from lifting π indefinitely is hitting a pair (f_j, c_j), that is (15) being violated for some j. In this case, we must do the HIT step of the algorithm. Also, (16) and (17) could stop us, and in those cases, we terminate. The algorithm also terminates if the f_j is hit at an increasing breakpoint.

Initially, every (f_j, c_j), $j \in J_1$, is above π; that is, $\pi(f_j) < c_j$, unless $c_j = 0$ since $\pi(f) = 0$ for every f, $0 \leq f \leq 1$. In the algorithm, once $\pi(f_j) = c_j$ and f_j has been made into a fixed breakpoint the pair (f_j, c_j) will be called a hit point. In addition, points (f_j, c_j) will be generated by the algorithm. Thus (f_j, c_j), $j \in J_1$ will be called original points. Initially, every (f_j, c_j), $j \in J_1$, is an original point and is an above point.

Let

$$c^+ = \min \{\frac{c_j}{a_{ij}} : j \in J_2^+\}, \qquad \text{and}$$

$$c^- = \min \{\frac{c_j}{-a_{ij}} : j \in J_2^-\}.$$

Then (16 and (17) become $\pi^+ < c^+$ and $\pi^- \leq c^-$.

The general steps of the algorithm are given below.

LIFT: For each above point (original or generated) (f_j, c_j), let $e_i \leq f_j < e_{i+1}$ for breakpoints e_i, e_{i+1}. Let $\delta_j = \alpha\delta_\ell + (1-\alpha)\delta_u$ where $\delta_\ell(\delta_u)$ is 0 or 1 depending on whether $e_i(e_{i+1})$ is fixed or increasing respectively and where

$$\alpha = \frac{e_{i+1} - f_j}{e_{i+1} - e_i}.$$

Let $\pi_j = \alpha\pi(e_i) + (1-\alpha)\pi(e_{i+1})$. Let $\delta_f = \min \{\frac{c_j - \pi_j}{\delta_j}$: all above points $(f_j, c_j)\}$. In the above, if $c_j - \pi_j = 0$, consider the ratio to be 0 even if $\delta_j = 0$. If $\delta_j = 0$ and $c_j - \pi_j > 0$, then the ratio is $+\infty$. Let $\delta^+ = c^+ e_1 - \pi(e_1)$ and $\delta^- = c^-(1-e_{L-1}) - \pi(e_{L-1})$, where e_{L-1} is the largest breakpoint $e_{L-1} < 1$. If either δ^+ or δ^- is less than or equal to δ_f, then change $\pi(e)$ to $\pi(e) + \min\{\delta^+, \delta^-\}$ for increasing breakpoints e and terminate. If $\delta_f < \delta^+$ and $\delta_f < \delta^-$, then the critical points (f_j, c_j) in defining δ_f become hit points. Change $\pi(e)$ to $\pi(e) + \delta_f$ for all increasing breakpoints e. If any hit point (f_j, c_j) has $f_j = e$ for some increasing breakpoint e, then terminate. Otherwise, go to the HIT step below.

HIT: Every hit point (f_j, c_j) is made into a fixed breakpoint. Each complimentary place $f_0 - f_j$ (modulo 1) becomes an increasing breakpoint.

GENERATE: For each new hit point (f_j, c_j), generate new (f_k, c_k) as specified below:

$$(f_k, c_k) = (f_j, c_j) + (e_i, \pi(e_i)),$$

for all fixed breakpoints $(e_i, \pi(e_i))$. However, if the new hit point (f_j, c_j) is a generated point, then the above sum is formed only for fixed breakpoints $(e_i, \pi(e_i))$ which were original points; that is, $(e_i, \pi(e_i)) = (f_\ell, c_\ell)$ for some $\ell \in J_1$. The new f_k is taken modulo 1 in the above sum. Return to LIFT.

The ratio $(c_j - \pi_j)/\delta_j$ formed in LIFT does not change as long as the interval containing f_j does not change. Whenever a new point is hit, it breaks an existing interval in two, and the new values of the ratio must be computed for all f_j in the two new intervals.

Each generated point (f_k, c_k) is a non-negative integer combination of original points (f_ℓ, c_ℓ) as can easily be seen by induction. Thus, for each generated point (f_k, c_k), there is a way to generate f_k from the original points at a cost of c_k. To show that the algorithm does solve the problem, we must show that the function π remains subadditive and that the algorithm terminates with the generated point (f_0, c_0) where $c_0 = \pi(f_0)$.

The algorithm terminates whenever $\pi^+ = c^+$, $\pi^- = c^-$ or the hit f_j is at an increasing breakpoint. Consider first the case $\pi^+ = c^+$. By (12), the slope at f_0 from below must also be π^+. Now the first breakpoint below f_0 must be fixed or else π^+ would not be increasing. That fixed breakpoint $(e, \pi(e))$ can be reached using some non-negative integer combination of original (f_j, c_j). From the breakpoint, s^+ can be used to reach $(u_0, \pi(f_0))$ where

$$\pi(f_0) = c^+(u_0 - e) + \pi(e).$$

Thus a solution $x^* \in T$ can be generated with

$$\pi(f_0) = \sum_{j \in J_N} c_j x_j^* \quad .$$

The case $\pi^- = c^-$ is similar using 5 .

Now, if the hit f_j is an increasing breakpoint, then its complementary point $f_0 - f_j$ is fixed and $(f_0, \pi(f_0))$ can be generated as the sum of two generated, hit points:

$$(f_j, \pi(f_j)) + (f_0 - f_j, \pi(f_0 - f_j)) .$$

3. Validity of the Algorithm

In view of Theorem 2 and the fact that (15), (16), and (17) are clearly satisfied, we need only show that our π produced by the algorithm is in $\Pi(f_0)$. Of the requirements (10) through (13), only (11), i.e. subadditivity, is not easy to show. Condition (12), $\pi(u) + \pi(f_0-u) = \pi(f_0)$, is true for breakpoints u because the breakpoints are paired so that whenever u is fixed, $f_0 - u$ is increasing. Thereafter, $\pi(u)$ remains the same and $\pi(f_0-u)$ increases by the same amounts as $\pi(f_0)$. The condition (12) is true for arbitrary u because π is a linear interpolation.

The proof of subadditivity for π is based on the following theorem.

Theorem 3: If π is a piecewise linear function satisfying all of the following:

$$\pi(u) \geq 0 \ , \ 0 \leq u \leq 1 \ , \quad \text{and} \quad \pi(0) = \pi(1) = 0 \ ;$$

$$\pi(u) + \pi(f_0-u) = \pi(f_0) \ , \quad \text{for} \ 0 \leq u \leq 1 \ ;$$

$$\pi(u) + \pi(v) \geq \pi(u+v), \quad \text{for convex breakpoints} \ u, v \ ;$$

then π is subadditive; that is, $\pi(u)+\pi(v) \geq \pi(u+v)$ for all u, v, $0 \leq u, v, \leq 1$.

A convex breakpoint of a piecewise linear function is a breakpoint such that the left slope is less than the right slope. In Figure 1, v is a convex breakpoint, and u+v is a concave breakpoint. The points 0 and 1 are considered to be convex breakpoints. In the algorithm, fixed break-points become convex and increasing breakpoints become concave breakpoints.

Proof of the Theorem: Suppose not; that is, suppose

$$\pi(u) + \pi(v) < \pi(u+v)$$

for some u, v. If both u and v are on linear sections of π, then for some $\alpha > 0$ either

$$\pi(u+\alpha) + \pi(v-\alpha) \leq \pi(u) + \pi(v) \text{ , or}$$

$$\pi(u-\alpha) + \pi(v+\alpha) \leq \pi(u) + \pi(v) \text{ .}$$

In both cases $u' + v' = u + v$, where $u' = u+\alpha$ (or $u - \alpha$) and $v' = v-\alpha$ (or $v + \alpha$). Hence, for one of these pairs u', v',

$$\pi(u') + \pi(v') < \pi(u' + v') \text{ .}$$

The α can be made large enough that one of u', v' reaches a breakpoint. If the breakpoint were a concave breakpoint, the inequality $\pi(u') + \pi(v') < \pi(u) + \pi(v)$ is just strengthened when α is increased. Hence, α can be increased until one of u', v' reaches a convex breakpoint, say v'.

Now, u' can be changed to either $u'' = u' + \beta$ or $u'' = u' - \beta$ for some $\beta > 0$ and either

$$\pi(u'+\beta)+\pi(v')-\pi(u'+v'+\beta) \leq \pi(u')+\pi(v')-\pi(u'+v'), \text{ or}$$

$$\pi(u'-\beta)+\pi(v')-\pi(u'+v'-\beta) \leq \pi(u')+\pi(v')-\pi(u+v') \text{ .}$$

Then, u'' can be moved until either u'' reaches a convex breakpoint or $u'' + v'$ reaches a concave breakpoint. In the former case, a contradiction is reached because u'' and v' are both convex breakpoints, and π was assumed subadditive then. In the latter case, the situation is as pictured in Figure 1. Let us drop the primes and consider the case where

$$\pi(u) + \pi(v) < \pi(u + v)$$

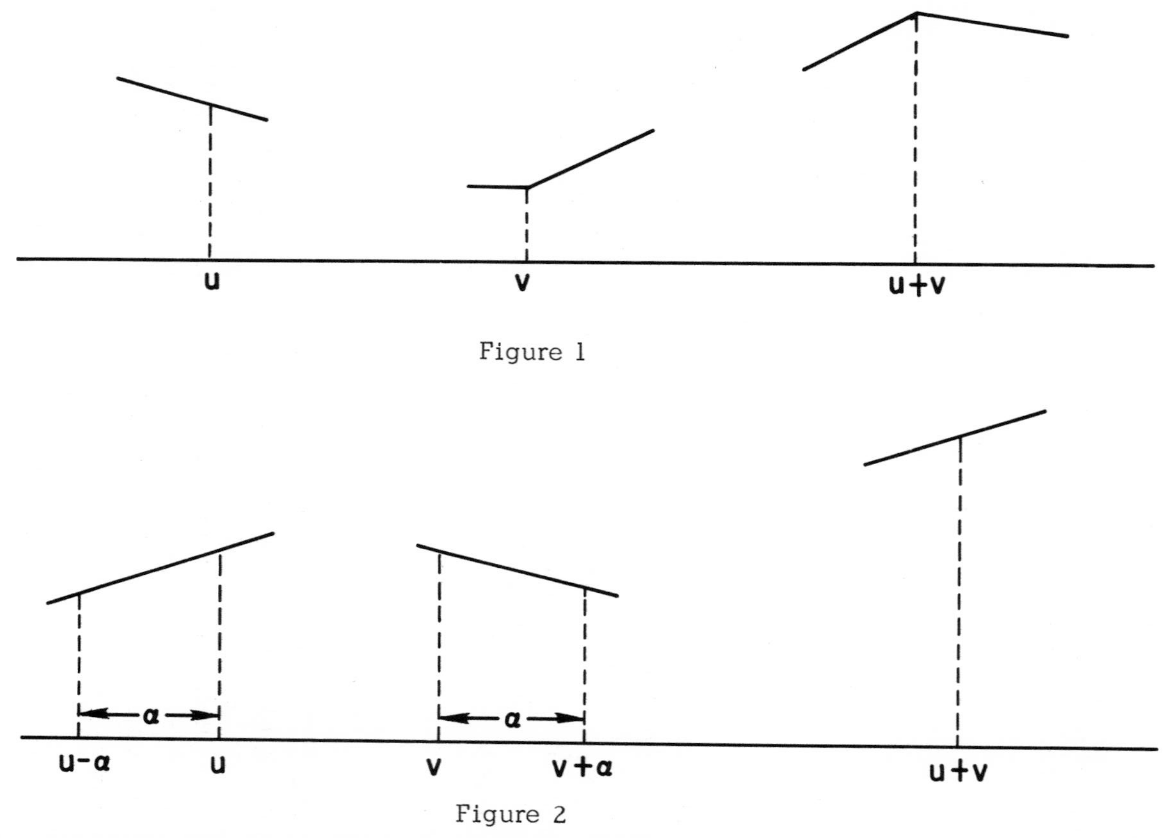

Figure 1

Figure 2

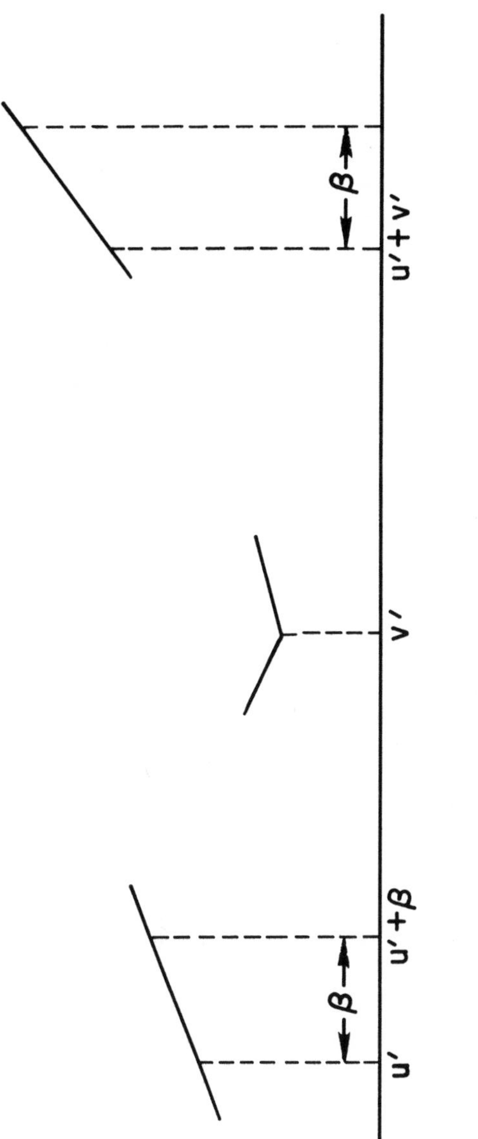

Figure 3

where v is a convex breakpoint and $u+v$ is a concave breakpoint.

If $f_0 = u+v$, a contradiction is reached because $\pi(u) + \pi(f_0-u) = \pi(f_0)$. If $u+v \neq f_0$, then $f_0 - (u+v)$ is a convex breakpoint by $\pi(w) + \pi(f_0-w) = \pi(f_0)$ for all w. Further,

$$\pi(u+v) + \pi(f_0 - (u+v)) = \pi(f_0) .$$

Substituting $\pi(u+v)$ from above gives

$$\pi(u) + \pi(v) < \pi(f_0) - \pi(f_0-(u+v)), \quad \text{or}$$

$$\pi(u) + \pi(v) + \pi(f_0 - (u+v)) < \pi(f_0) .$$

Since both v and $f_0 - (u+v)$ are convex breakpoints, $\pi(v) + \pi(f_0-(u+v)) \geq \pi(f_0-u)$, so

$$\pi(u) + \pi(f_0-u) < \pi(f_0) .$$

Thus, a contradiction is reached since the equality was assumed for convex breakpoints u.

The proof that the π in the algorithm remains subadditive would be complete if in the GENERATE step the sum $(f_j, c_j) + (e_i, \pi(e_i))$ had been formed for all fixed breakpoints e_i. If it had been, then $\pi(e_i) + \pi(e_j)$ would be an upper bound on $\pi(e_i + e_j)$ for all convex breakpoints e_i since all convex breakpoints are fixed. However, the sum was only formed for original fixed breakpoints e_i whenever (f_j, c_j) was a generated point. The proof will be by induction on the LIFT step of the algorithm. Initially $\pi(u) = 0$ and the only fixed breakpoints are 0 and 1. As long as subadditivity holds for all pairs of fixed breakpoints, Theorem 3 proves that π is subadditive. Suppose during LIFT subadditivity becomes violated for the two fixed breakpoints e_i and e_j. Then there is a point at which

$$\pi(e_i) + \pi(e_j) = \pi(e_i + e_j)$$

but further lifting of π would increase $\pi(e_i + e_j)$ causing violation of subadditivity. We will show that LIFT would stop here because some (f_k, c_k) is already hit.

We know that

$$e_i \equiv \sum_{\ell \in J_1} \lambda_{i\ell} f_\ell, \quad \pi(e_i) = \sum_{\ell \in J_1} \lambda_{i\ell} c_\ell , \qquad \text{and}$$

$$e_j \equiv \sum_{\ell \in J_1} \lambda_{j\ell} f_\ell, \quad \pi(e_j) = \sum_{\ell \in J_1} \lambda_{j\ell} c_\ell ,$$

where $\lambda_{i\ell}$ and $\lambda_{j\ell}$ are some non-negative integers, by the way points are generated. Furthermore, if any $\lambda_{i\ell} > 0$ or $\lambda_{j\ell} > 0$, then $\pi(f_\ell) = c_\ell$; that is, (f_ℓ, c_ℓ) is an original fixed breakpoint.

By $\pi(e_i) + \pi(e_j) = \pi(e_i + e_j)$,

$$\pi(e_i + e_j) = \sum_{\ell \in J_1} \lambda_{i\ell} c_\ell + \sum_{\ell \in J_1} \lambda_{j\ell} c_\ell , \qquad \text{or}$$

$$\pi\Big(\sum_{\ell \in J_1} (\lambda_{i\ell} + \lambda_{j\ell}) f_\ell \Big) = \sum_{\ell \in J_1} (\lambda_{i\ell} + \lambda_{j\ell}) c_\ell .$$

Since $\lambda_{i\ell} + \lambda_{j\ell} > 0$ implies $\pi(f_\ell) = c_\ell$, we can write

$$\pi\Big(\sum_{\ell \in J_1} (\lambda_{i\ell} + \lambda_{j\ell}) f_\ell \Big) = \sum_{\ell \in J_1} (\lambda_{i\ell} + \lambda_{j\ell})\, \pi(f_\ell) .$$

This π has been assumed to be subadditive, but any further increase would violate subadditivity. Hence, for any λ_ℓ , $0 \le \lambda_\ell \le \lambda_{i\ell} + \lambda_{j\ell}$,

$$\pi\Big(\sum_{\ell \in J_1} \lambda_\ell f_\ell \Big) \ge \sum_{\ell \in J_1} \lambda_\ell\, \pi(f_\ell) , \qquad \text{and}$$

$$\pi\Big(\sum_{\ell \in J_1} (\lambda_{i\ell} + \lambda_{j\ell} - \lambda_\ell) f_\ell \Big) \ge \sum_{\ell \in J_1} (\lambda_{i\ell} + \lambda_{j\ell} - \lambda_\ell) \pi(f_\ell) .$$

Adding and using subadditivity once more on the left-hand side gives

$$\pi\Big(\sum_{\ell \in J_1} (\lambda_{i\ell} + \lambda_{j\ell}) f_\ell\Big) \geq \sum_{\ell \in J_1} (\lambda_{i\ell} + \lambda_{j\ell}) \pi(f_\ell) .$$

But this inequality holds with equality, and so the above two must as well; that is, for any $0 \leq \lambda_\ell \leq \lambda_{i\ell} + \lambda_{j\ell}$,

$$\pi\Big(\sum_{\ell \in J_1} \lambda_\ell f_\ell\Big) = \sum_{\ell \in J_1} \lambda_\ell \pi(f_\ell) .$$

In particular, let $\lambda_\ell = \lambda_{i\ell}$ except for one $\ell = k$ for which $\lambda_{jk} > 0$ let $\lambda_k = \lambda_{ik} + 1$. Then, $\Sigma \lambda_\ell f_\ell = e_i + f_k$ and

$$\begin{aligned} \pi(e_i + f_k) &= \pi(e_i) + \pi(f_k) \\ &= \pi(e_i) + c_k . \end{aligned}$$

Since e_i is a fixed breakpoint and f_k is an original fixed breakpoint, we would have generated the point $(e_i + f_k, \pi(e_i) + e_k)$. Since $\pi(e_i + f_k) = \pi(e_i) + c_k$, in LIFT $\delta_f = 0$ and we would stop because of hitting this point.

4. Cyclic Groups

Termination of the algorithm is based on rationality of the f_j, f_0. In practice, a tolerence $\varepsilon > 0$ will achieve the same result. Let us consider now the case $f_j = g_j/D$, $j \in J_1$, and for $f_0 = g_0/D$. If we are willing to store vectors of size D, the required work can be reduced and bounded. The algorithm will first be reformulated.

Suppose no c^+ or c^- is present to limit the slopes.

For each $i = 1, \ldots, D$, initially let $\Delta(i) = \infty$ except that

$$\Delta(g_j) = \begin{cases} \dfrac{f_0 c_j}{f_j}, & g_j \leq g_0 , \\[2ex] \dfrac{(1-f_0)c_j}{1-f_j}, & g_j > g_0 , \end{cases}$$

for $j = 1, \ldots, N_1$. The candidate set is, initially,

$$C_S = \{f_j : j = 1, \ldots, N_1\} .$$

For each f_j, $j \in J_1$, let $\gamma(f_j) = c_j$. For other $f \neq$ any f_j, let $\gamma(f) = \infty$. The set I of increasing points is initially $\{f_0\}$, and the set H of fixed points is initially $\{0, 1\}$. The set E of breakpoints is $H \cup I$ and is initially $\{0, f_0, 1\}$. Let $\pi(0) = \pi(f_0) = \pi(1) = 0$. Let $e_1 = 0$ and $e_2 = f_0$, and

$$\delta_1 = \min \{\Delta(g) : f \in C_s \text{ and } 0 \leq f < g_0\}$$

$$\delta_2 = \min \{\Delta(g) : f \in C_s \text{ and } g_0 \leq f < D\} .$$

The algorithm is restated below.

LIFT: Let

$$\delta^* = \min \{\delta_i : e_i \text{ is a breakpoint}\} .$$

Each i such that $\delta_i = \delta^*$ is critical, and for a criticial i, every f in C_S such that $e_i \leq f < e_{i+1}$ and $\Delta(f) = \delta_i$ is a new hit point.. Remove all these hit points from C_S.

HIT: Adjoin every new hit point h to the set H of fixed points. The complimentary element f_0-h (modulo D) be - comes an increasing point and is put in the set I. If any new hit point is equal to f_0 or to any other point in I, then terminate. Otherwise, go to generate.

GENERATE: For each new hit point $(h, \gamma(h))$ (recall that initially $\gamma(g_j) = c_j$), let

$$\gamma(g+h) = \min \{\gamma(g+h), \ \gamma(g) + \gamma(h)\}$$

where g is either

(i) all $g \in H$, including new hit points, when h is some g_j and $\gamma(h) = c_j$;

(ii) all $g = g_j$ for all j for which $g_j \in H$ and $\gamma(g_j) = c_j$. Adjoin $g+h$ to the candidate set C_s if $\gamma(g+h)$ was ∞ before this step.

<u>RECOMPUTE Δ :</u> For certain g, $\Delta(g)$ must be recomputed giving rise to new values δ_i for some i. Suppose first that the elements of $E = \{0 = e_1, e_2, \ldots, e_k = 1\}$ are ordered so that $e_1 < e_2 < \ldots < e_i < e_{i+1}, \ldots$. Certain of the e_i are new; either new hit points or new increasing points. In addition, GENERATE lowered some values of $\gamma(g)$.

For a g such that $\gamma(g)$ was lowered in GENERATE, $\Delta(g)$ must be recomputed. In addition, for any $g \in C_s$ such that for some $e_\ell \in E$, $e_\ell \leq g < e_{\ell+1}$ and either e_ℓ or $e_{\ell+1}$ is a new member of E, then $\Delta(g)$ must be recomputed. The formula for $\Delta(g)$ is given below. First of all, if $e_\ell \in H$, then $\pi(e_\ell) = \gamma(e_\ell)$ and if $e_\ell \in I$, then $\pi(e_\ell) = \delta^* - \gamma(g_0 - e_\ell)$ where δ^* is from LIFT and $g_0 - e_\ell$ is taken modulo D so $g_0 - e_\ell \in H$. Similarly for $e_{\ell+1}$. Then,

$$\pi(g) = \pi(e_\ell)\frac{e_{\ell+1}-g}{e_{\ell+1}-e_\ell} + \pi(e_{\ell+1})\frac{g-e_\ell}{e_{\ell+1}-e_\ell}, \quad \text{and}$$

$$d(g) = d(e_\ell)\frac{e_{\ell+1}-g}{e_{\ell+1}-e_\ell} + d(e_{\ell+1})\frac{g-e_\ell}{e_{\ell+1}-e_\ell},$$

where $d(e_\ell)$ is 1 if $e_\ell \in I$ and 0 is $e_\ell \in H$, and similarly for $d(e_{\ell+1})$. Now

$$\Delta(g) = \delta^* + \frac{\gamma(g) - \pi(g)}{d(g)} .$$

If $d(g) = 0$, then $\Delta(g) = \infty$.

If a g for which $\gamma(g)$ was lowered falls in an interval $e_\ell \leq g < e_{\ell+1}$ for which neither e_ℓ nor $e_{\ell+1}$ is a new member of E, then δ_ℓ is replaced by the minimum of δ_ℓ and $\Delta(g)$. However, for an interval $(e_\ell, e_{\ell+1})$ where either e_ℓ or $e_{\ell+1}$ is a new member of E, all candidates g, $e_\ell \leq g < e_{\ell+1}$ must have $\Delta(g)$ recomputed and δ_ℓ is then the minimum of all such $\Delta(g)$.

Return now to LIFT.

In this version of the algorithm, δ^* is the present value of $\pi(g_0)$, and $\Delta(g)$ is the value of $\pi(g_0)$ for which, with the present $E = H \cup I$, $\pi(g) = \gamma(g)$. For each $g \in C_s$ or $g \in H$,

$$g = \sum_{j=1}^{N_1} \lambda_{gj} g_j , \qquad \text{and}$$

$$\gamma(g) = \sum_{j=1}^{N_1} \lambda_{gj} c_j ,$$

for some non-negative integers λ_{gj}.

For each $h \in H$, $\pi(h) = \gamma(h)$, and for each $g \in I$, $\pi(g) = \pi(g_0) - \pi(g_0 - g)$, where $g_0 - f$ is taken modulo D and is in H. If $\gamma(g) = \pi(g)$ for some $g \in I$, then the algorithm terminates. Let $h = g_0 - g$ (modulo D), so $h \in H$ and $\pi(h) = \pi(g_0) - \pi(g) = \sum_{j=1}^{N_1} \lambda_{hj} c_j$ for some non-negative integers λ_{hj} for which $h = \Sigma \lambda_{hj} g_j$. Since $g+h = g_0$ (modulo D),

$$g_0 = \sum_{j=1}^{N_1} (\lambda_{hj} + \lambda_{gj}) g_j , \qquad \text{and}$$

$$\pi(g_0) = \pi(g) + \pi(h) = \gamma(g) + \gamma(h)$$

$$= \sum_{j=1}^{N_1} (\lambda_{hj} + \lambda_{gj}) c_j .$$

Hence, $\lambda_{hj} + \lambda_{gj}$, $j = 1, \ldots, N_1$, is a solution to the group problem such that (14) holds.

The proof that our $\pi \in \Pi(f_0)$ is the same as in Section 2 and will not be repeated.

We now turn to consideration of upper bounds on the work required for the algorithm to produce a solution. The worst case is where each iteration produces one new hit point h, and, henceforth, that case will be considered.

The maximum number of iterations possible is $D/2$, since each iteration puts two more elements in E.

The work required in HIT per iteration is a constant. If for each $g \in C_S$, we record the breakpoints $e_\ell \in E$ and $e_{\ell+1}$ such that $e_\ell \le g < e_{\ell+1}$, then reordering E is also just a constant amount of work. We need also to store, for each breakpoint e, its complimentary breakpoint $g_0 - e$. Updating the two breakpoints on either side of the remaining $g \in C_S$ can be done during RECOMPUTE Δ where such information is already available.

In GENERATE, the total work done during all iterations is at most $N_1 \times D/2$ additions and the same number of comparisons. The number of each operation per iteration is $|H|$ if the new hit point is a g_j and is at most N_1 otherwise. However, the total is at most one addition and one comparison for each pair g_j, h where $j = 1, \ldots, N_1$ and $h \in H$.

The work done in LIFT per iteration can be bounded above by $|E|$ comparisons and a constant amount of work to find the new hit points. The total amount of work here is thus bounded by a constant times D^2. In practice, we can improve on the actual amount of work done by using a binary sort of δ_i [5]. The reason is that usually not all of the δ_i will change, and the work to update the binary sort will usually be less than $|E|$. Even in the worse case, it is only of order $|E|$. The number of δ_i which change is at most four for the four intervals adjacent to the new fixed point and increasing point. In addition, the δ_i may change in intervals which include some $g+h$ from the GENERATE step. Except for the (at most N_1) iterations when the new hit point is some g_j, the number of δ_i which change is at most N_1+4. For a binary sort, if N_1+4 numbers change, then at most $(N_1+4) \log |E|$ operations are required to update the binary sort. When N_1+4 is less than $|E|/\log|E|$ this number is less than $|E|$.

In RECOMPUTE Δ, a constant amount of work is required for each g for which $\Delta(g)$ must be recomputed. The number of such g could be, at worst, the entire candidate set. If we add up $|C_S|$ over all iterations, the sum could be of order D^2. Thus, the overall bounds remains D^2.

In Section 6, some actual counts for five problems will be given.

5. Cutting Planes

As shown by Theorem 1, for $\pi \in \Pi(f_0)$, the inequality (18), i.e.

$$\sum_{j \in J_1} \pi(f_j)x_j + \sum_{j \in J_2^+} \pi^+ \bar{a}_{ij}x_j - \sum_{j \in J_2^-} \pi^- \bar{a}_{ij}x_j \geq \pi(f_0)$$

is a cutting plane, or valid inequality, for the congruence

$$\sum_{j \in N} \bar{a}_{ij}x_j \equiv f_0, \; x_j \geq 0 \; , \; x_j \text{ integer for } j \in J_1 \; , \text{ where}$$

$$f_j = F(\bar{a}_{ij}) \; , \text{ the fractional part of } \bar{a}_{ij} \text{ for } j \in J_1 \; .$$

Consider, for example, the congruence

$$1\frac{1}{2}x_1 - \frac{1}{3}x_2 + \frac{1}{10}x_3 + 1\frac{1}{6}x_4 - \frac{1}{9}x_5 \equiv \frac{1}{3}$$

where x_1 and x_2 are required to be integer. Then, $f_1 = \frac{1}{2}$ and $f_2 = \frac{2}{3}$, $J_2^+ = \{3,4\}$ and $J_2^- = \{5\}$. Gomory's mixed integer cut ([1] page 528) is given by

$$\pi(u) = \begin{cases} \dfrac{u}{\frac{1}{3}} = 3u \; , & 0 \leq u \leq \frac{1}{3} \; , \\[2ex] \dfrac{1-u}{1-\frac{1}{3}} = \frac{3}{2}(1-u), & \frac{1}{3} < u \leq 1 \; , \end{cases}$$

$$\pi^+ = 1/(\tfrac{1}{3}) = 3 \; , \quad \pi^- = 1/(1 - \tfrac{1}{3}) = \tfrac{3}{2} \; .$$

The inequality given by this π is

$$\frac{3}{4}x_1 + \frac{1}{2}x_2 + \frac{3}{10}x_3 + 3\frac{1}{2}x_4 + \frac{1}{6}x_5 \geq 1 \; .$$

This π does belong to $\Pi(f_0)$ where $f_0 = \frac{1}{3}$ here.

Let us take as objective function to minimize

$$z = x_1 + x_2 + 2x_3 + 8x_4 + x_5 \; .$$

To satisfy the above inequality with non-negative variables and minimize z, let $x_1 = 4/3$ and $z = 4/3$. Rescaling the inequality gives

$$x_1 + \frac{2}{3}x_2 + \frac{2}{5}x_3 + 4\frac{2}{3}x_4 + \frac{2}{9}x_5 \geq \frac{4}{3}.$$

Now, (15), (16), and (17) are satisfied. Figure 4 illustrates the rescaled function $\frac{4}{3}\pi$. The two x's

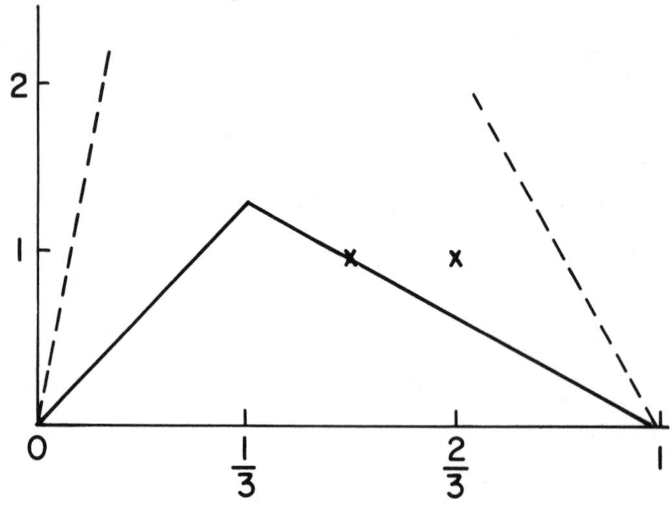

Figure 4

mark the points $(f_1, c_1) = (\frac{1}{2}, 1)$ and $(f_2, c_2) = (\frac{2}{3}, 1)$. The dotted lines indicate the limits on the slopes π^+ and π^- at the ends of the interval $[0,1]$. These restrictions are (16) and (17) and here are

$$\pi^+ \leq \frac{c_3}{a_3} = \frac{2}{\frac{1}{10}} = 20\ ,$$

$$\pi^- \leq \frac{c_4}{-a_4} = \frac{8}{\frac{7}{6}} = \frac{48}{7} = 6\frac{6}{7}\ ,$$

$$\pi^- \leq \frac{c_4}{-a_4} = \frac{1}{\frac{1}{9}} = 9 .$$

The algorithm given here begins with this π after one iteration and fixes $\pi(0) =)$, $\pi(1/2) = 1$, $\pi(1) = 0$, while letting $\pi(1/3) = 4/3 + \Delta$, $\pi(\frac{5}{6} \equiv \frac{1}{3} - \frac{1}{2}) = 1/3 + \Delta$. Figure 5 illustrates this change.

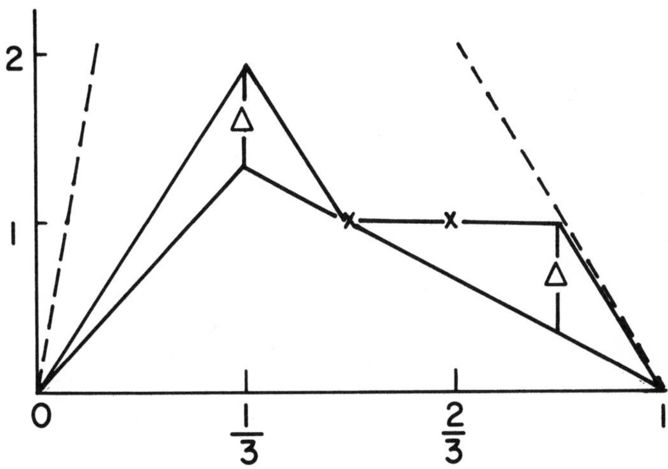

Figure 5

When $\Delta = 2/3$, $\pi(2/3) = 1 = c_2$. The slope π^+ is then 6 which is less than c_3/a_3. The x at (1, 2) results from (1/2, 1) + (1/2, 1) = (1, 2). Now, when (2/3, 1) + (2/3, 1) is formed, we get (1/3, 2). But $\pi(1/3) = 2$, so the algorithm terminates since 1/3 is an increasing breakpoint. The inequality given is

$$x_1 + x_2 + \frac{6}{10} x_3 + 7 x_4 + \frac{2}{3} x_5 \geq 2 .$$

This inequality does not imply the previous inequality given by the mixed integer cut, but it obviously costs more to satisfy with non-negative variables than the mixed integer cut.

6. Shortest Paths

The group problem (even when the group is not cyclic but more generally a direct product of cyclic groups) can be formulated as a shortest path problem (see [2], Appendix 2). The algorithm given here is a method of solving that shortest path problem and can be compared to other methods. A comparison with Hu's ([4], page 348) adaptation of the Dykstra algorithm will be made by means of an example. Some computational experience is also reported in order to make comparisons with the upper bounds on work required.

The group problem for a cyclic group of order D is a shortest path problem on a network having nodes $0, 1, \ldots, D-1$ and arcs $(g, g+g_j)$, for every $g = 0, 1, \ldots, D-1$ and every $j = 1, \ldots, N_1$, with length (or cost) c_j. This algorithm is outlined below.

Let $d_g = \infty$ except $d_0 = 0$. Let $C_s = \{0\}$,

General Step: For $h \in C_s$ with $d_h = \min\{d_g : g \in C_s\}$, remove h from C_s. If $h = g_0$, terminate. Otherwise, for each $j = 1, \ldots, N_1$ let $g = h + g_j$ and replace d_g by

$$\min\{d_g, d_h + c_j\}.$$

If any such d_g was ∞, place g in C_s. Repeat the general step.

The work in finding the min can be reduced by using a binary sort [5], but for purposes of comparison we will count the number of iterations and the total number of d_g lowered during the course algorithm. The work in this algorithm is N_1 times the number of iterations plus the work in finding the minimum d_g for $g \in C_s$. This latter work depends on the number of iterations and the number of times d_g is lowered during the algorithm.

There are several similarities between this algorithm and ours. For one, they are both dual methods in that they both give a lower bound on the optimum solution. In Hu's algorithm, the value d_h is a lower bound.

Both algorithms can be used to generate a cutting

plane. Hu's algorithm gives a valid inequality

$$\sum_{j=1}^{N_1} \min \{d_h, d_{g_j}\} x_j \geq d_h$$

at each iteration. The reason this inequality is valid is that the function

$$\pi(u) = \begin{cases} \min\{d_h, d_g\}, & \text{if } u = \frac{g}{D}, \\ d_h, & \text{otherwise} \end{cases}$$

is a subadditive function on $[0,1]$. Figure 6 illustrates this function for the problem:

$$x_1, x_2, x_3 \geq 0 \text{ and integer,}$$

$$3x_1 + 29x_2 + 91x_3 \equiv 52 \text{ (modulo 100)}$$

$$x_1 + 2x_2 + 3x_3 = z(\min).$$

Figure 6 shows π after 28 iterations for which $h = 3, 6, 29, 9, 32, 91, 12, 35, 58, 94, 15, 20, 38, 61, 97, 18, 23, 41, 64, 82, 87, 21, 26, 44, 49, 67, 85, 90$. The function π has value 8 except for the x places corresponding to points (u, d) where $\pi(u) = d$. At this point, 52 is hit with $d_{52} = 8$. An optimal solution is $x_1 = 1$, $x_2 = 2$, and $x_3 = 1$. Clearly $\pi(u) + \pi(v) \geq \pi(u+v)$ if either u or v has $\pi(u) = d_h$ or $\pi(v) = d_h$. If both $\pi(u) < d_h$ and $\pi(v) < d_h$, then $\pi(u) + \pi(v) = d_g + d_f \geq \pi(u+v)$ because the d_g and d_f are distances in a shortest path method. The proof is much the same as in Section 3 for two convex breakpoints u and v.

The algorithm in Section 4 gives the function shown in Figure 7 after five iterations hitting 29, 58, 91, 20, 49. At this point, $\pi(.52) = 8$ and the problem is solved. While it is true that this method requires more work per iteration than Hu's, the bound on the total work grows at the same rate, D^2, and our method usually moves faster.

Five randomly generated problems were solved using both methods. The problems involved 10 group elements from

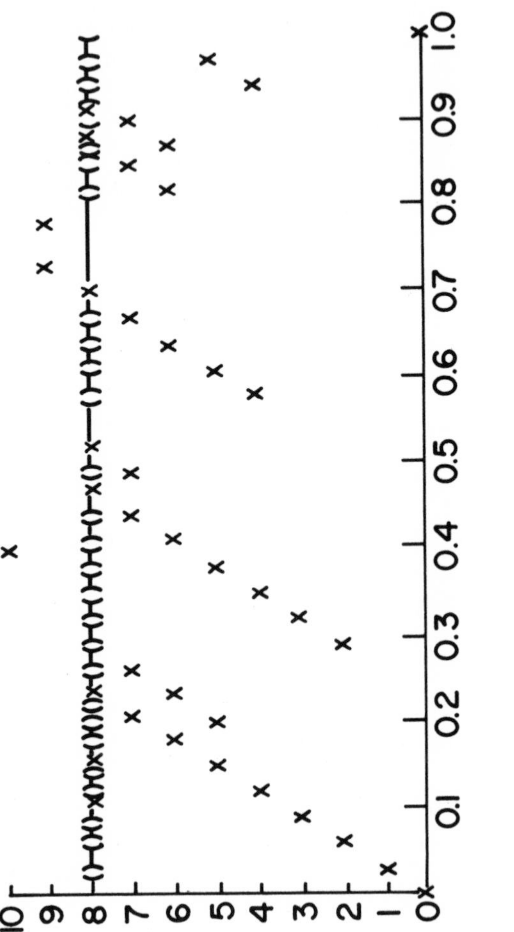

Figure 6

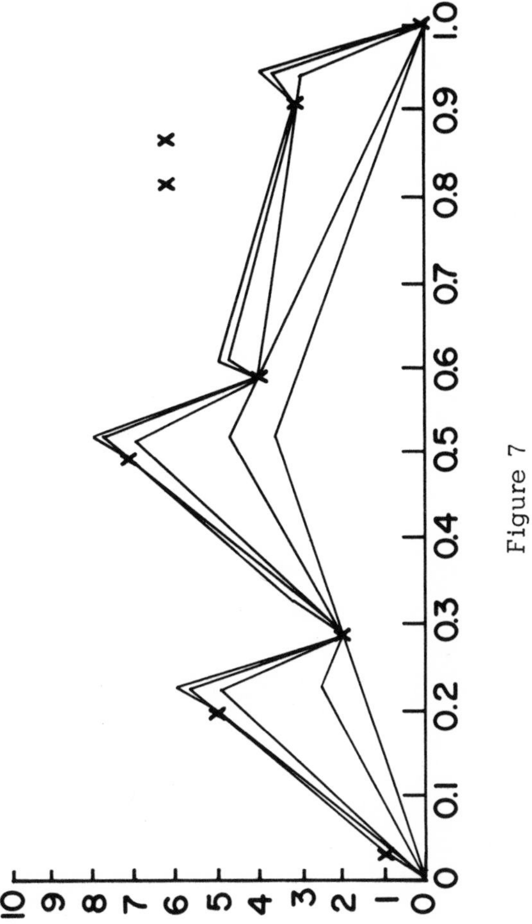

Figure 7

a uniform distribution of the integers 1, 2, ..., 1000. Figure 8 summarizes the results. The column "iter" is the number of iterations, "max set" is the maximum size of the candidate

Method	Order	Iter	Max Set	Comparisons	Reductions	Δ
Hu	23	5	18	50	26	
Lift	23	2	9	3	11	15
Hu	41	20	30	200	62	
Lift	41	6	12	18	19	40
Hu	56	12	35	120	49	
Lift	56	3	9	7	10	57
Hu	82	72	61	720	143	
Lift	82	39	9	48	40	73
Hu	100	85	65	850	200	
Lift	100	13	27	50	45	81

Figure 8

set, and "comparisons" is the number of times the operation $\min \{d_{g+h}, d_g + c_h\}$ was performed. This operation requires two additions as well as a comparison. The column "reductions" is the number of times the min was given by $d_g + c_h$. It also includes the 10 initial reductions, which accounts for there being more reductions than comparisons in some cases. The column "Δ" given the number of $\Delta(g)$ computed in the method given in Section 4.

The method given here can be seen to shift most of the work from the shortest path type operations to the arithmetic work of computing $\Delta(g)$. Of course, there is also some overhead because the algorithm here is more complicated and the program to execute it is longer.

REFERENCES

1. Dantzig, G. B., Linear Programming and Extensions, Princeton University Press, Princeton, N. J., 1962.

2. Gomory, R. E., "Some Polyhedra Related to Combinatorial Problems", Linear Algebra and Its Applications, 2 (1969), pp. 451-558.

3. Gomory, R. E. and E. L. Johnson, "Some Continuous Functions Related to Corner Polyhedra", RC 3311, IBM Research, Yorktown Heights, N. Y., February 23, 1971.

4. Hu, T. C., Integer Programming and Network Flows, Addison Wesley, Reading, Mass., 1969.

5. Johnson, E. L., "On Shortest Paths and Sorting", IBM RC 3691, IBM Research, Yorktown Heights, N. Y., 1972.

Mathematical Sciences Department
IBM Watson Research Center
Yorktown Heights, New York 10598

Use of Cyclic Group Methods in Branch and Bound

HARLAN P. CROWDER AND ELLIS L. JOHNSON

This paper reports on some computational experience using the algorithm from "Cyclic Groups, Cutting Planes, and Shortest Paths" to solve integer programs within a branch and bound algorithm. Our use of these cutting planes to provide bounds is very much along the lines of Tomlin's [7] use of Gomory's mixed integer cut. In fact, that cut is what our cutting plane algorithm gives after one iteration so is included as one case of our method. The difference here is that we also test using more work to generate cuts in hopes that better bounds and better choices in the branch and bound algorithm will result in smaller total running times. This work follows earlier work [5].

To begin, the basic branch and bound algorithm used is the Dakin [3] variant of the Land and Doig [6] algorithm. For a fuller discussion see Beale [1] and Benichou et al [2]. Initially, the linear program obtained by dropping all of the integer restrictions is solved as a linear program. The algorithm has two general steps:

<u>Step 1:</u> Choose some linear program and solve it as a linear program;

<u>Step 2:</u> Choose some variable x_{k_i} which is required to be integer-valued but whose value $x^0_{k_i}$ is not integer. Then, create two new linear programs:

a down l. p. defined by requiring $x_{k_i} \leq \lfloor x^0_{k_i} \rfloor$
and
an up l. p. defined by requiring $x_{k_i} \geq \lceil x^0_{k_i} \rceil$.

Here $\lfloor x \rfloor$ means the largest integer less than or equal to x and $\lceil x \rceil$ is the smallest integer greater than or equal to x .

In step 1 there are four possible cases:

(a) the linear program has no feasible solution - go back to step 1 ; i. e. choose another linear program to solve;

(b) the objective value for the solved linear program is worse than a previously found integer solution - go back to step 1 ;

(c) the linear programming optimal solution has no integer variables at non-integer values - compare to the best integer solution found so far and record if better than the previous best;

(d) otherwise go to step 2 with a linear programming, but non-integer, solution.

There are two choices which must be made: the linear program to solve in step 1 and the variable to branch on in step 2. Once a variable in step 2 is chosen creating two new linear programs, we always choose the next linear program in step 1 to be one of these two linear programs. Whenever we return to step 1 from one of the three cases (a), (b), (c), we have not just created two new linear programs, and the choice of linear program is completely open. Each such choice begins a new major cycle. The first major cycle consists of solving the linear program, branching on some variable, solving one of these two linear programs, and continuing until the linear program chosen is either infeasible or has an integer optimum solution.

In tables 2 through 6, the column "major" refers to the number of major cycles required to solve the integer program (including proof of optimality). The column "no. lp" is the total number of linear programs solved, and the column "iter" is the total number of linear programming iterations (basis changes) required to solve the problem not counting the iterations to solve the continuous problem (the initial linear program). The times are in seconds of execution for our FORTRAN Code on an IBM 360/91.

Table 1 shows five problems tested. The first is a small, pure integer problem, the next two are zero-one problems, and the last two are small, mixed integer problems. The column "iter to lp" refers to the iterations required to solve the linear program obtained by relaxing the integrality restrictions.

The first method shown in tables 2 through 6 is "priority" which simply orders the integer variables in terms of increasing cost coefficients c_j (for a minimization problem) and branches on the first integer variable, in this order, which is not at an integer value. The linear program solved is then the down linear program obtained by rounding down. When returning to step 1, the linear program chosen is the one with highest value of the objective function. The linear programs are solved by the dual simplex method which restores feasibility after branching on a variable.

All of the other methods tested use <u>down penalties</u> p_{Dk} and <u>up penalties</u> p_{Uk} which are bounds applicable to the down linear program and the up linear program resulting from branching on the basic variable x_k. The penalties are such that any integer solution on the down branch, for example, has objective value no smaller than z_L+p_{Dk} where z_L is the linear programming solution at that point.

These penalties are used for selecting the variable to branch on. The tests used the selection rule: branch on variable x_k such that

$$\min\{p_{Dk}, p_{Uk}\} + \frac{1}{2}\max\{p_{Dk}, p_{Uk}\}$$

is maximized. Our experience is that it is important to choose a variable with a large value of $\min\{p_{Dk}, p_{Uk}\}$.

The penalties are also used for estimation [2] to choose the linear program. When we branch on the previously determined variable, say x_ℓ, the <u>estimate</u> of the down linear program is

$$z_L + p_{D\ell} + \sum_{k \neq \ell} \min \{p_{Dk}, p_{Uk}\} ,$$

and the estimate of the up linear program is

$$z_L + p_{U\ell} + \sum_{k \neq \ell} \min \{p_{Dk}, p_{Uk}\} .$$

We then choose the linear program with smallest estimate. Within a major cycle, this choice is between the down linear program and the up linear program and amounts to choosing the down one if $p_{D\ell}$ is smaller than $p_{U\ell}$.

All that remains is to describe the seven different ways in which the penalties are generated. The first [4] is called "lp pen" in the tables and consists of looking at the next iteration of the dual simplex algorithm after branching on some variable x_k. In more detail, suppose the ith basic variable is an integer variable at a non-integer value. The ith row of the updated tableau is

$$x_{ki} + \sum_{j \in J_N} \bar{a}_{ij} x_j = \bar{b}_i$$

where J_N is the index set of non-basic variables. If the x_j also have restrictions $\ell_j \leq x_j \leq u_j$, making the substitution $x'_j = x_j - \ell_j$ or $x'_j = u_j - x_j$, $j \in J_N$, depending on whether the basic solution x^0 has $x^0_j = \ell_j$ or $x^0_j = u_j$, gives

$$x_{k_i} + \sum_{j \in J_N} \alpha_j x'_j = \alpha_0$$

$$\sum_{j \in J_N} \gamma_j x'_j = z - z_L$$

where α_0 is the basic value of x_{k_i} and

$$\alpha_j = \begin{cases} \bar{a}_{ij}, & x^0_j = \ell_j \\ -\bar{a}_{ij}, & x^0_j = u_j \end{cases} ,$$

$$\gamma_j = \begin{cases} \bar{c}_j \,, & x_j^0 = \ell_j \\ -\bar{c}_j, & x_j^0 = u_j \,. \end{cases}$$

The down linear program restricts $x_{k_i} \leq \lfloor \alpha_0 \rfloor$, or $\Sigma \alpha_j x_j' \geq \alpha_0 - \lfloor \alpha_0 \rfloor$. Requiring this inequality with the objective function $z = z_L + \Sigma \gamma_j x_j'$ gives an optimum z of $z_L + p_{Dk}$ where

$$p_{Dk_i} = \min \{\frac{\gamma_j}{\alpha_j}(\alpha_0 - \lfloor \alpha_0 \rfloor) : j \in J_N \text{ and } \alpha_j > 0\} \,.$$

The up penalty is similarly the cost of requiring $x_{k_i} \geq \lceil \alpha_0 \rceil$, or $\Sigma(-\alpha_j)x_j' \geq \lceil \alpha_0 \rceil - \alpha_0$. Thus,

$$P_{Uk_i} = \min \{\frac{\alpha_j}{-\alpha_j}(\lceil \alpha_0 \rceil - \alpha_0) : j \in J_N \text{ and } \alpha_j > 0\} \,.$$

Tomlin [7] uses Gomory's mixed-integer cut in two ways. The first is to let p_{Ik_i} be the minimum value of $\Sigma \gamma_j x_j'$ subject to $x_j \geq 0$ and $\Sigma \pi_j x_j' \geq 1$ where this latter inequality is the mixed integer cut based on row i. That is, for $j \in J_N$,

$$\pi_j = \begin{cases} \dfrac{\alpha_j}{F(\alpha_0)}, & \alpha_j \geq 0 \text{ and } x_j \text{ continuous} \\[2ex] \dfrac{-\alpha_j}{1-F(\alpha_0)}, & \alpha_j < 0 \text{ and } x_j \text{ continuous} \\[2ex] \dfrac{F(a_j)}{F(\alpha_0)}, & F(\alpha_j) \leq F(\alpha_0) \text{ and } x_j \text{ integer} \\[2ex] \dfrac{1-F(\alpha_j)}{1-F(\alpha_0)}, & F(\alpha_0) < F(\alpha_j) \text{ and } x_j \text{ integer} \,. \end{cases}$$

This cut is based on the congruence

$$\sum_{j \in J_N} \alpha_j x_j \equiv \alpha_0 \quad (\text{modulo } 1) .$$

This penalty p_{Ik_i} can be used to strengthen the previous penalties p_{Dk_i} and p_{Uk_i} to p'_{Dk_i} and p'_{Uk_i} given by:

$$p'_{Dk_i} = \max\{p_{Ik_i}, p_{Dk_i}\}$$

$$p'_{Uk_i} = \max\{p_{Ik_i}, p_{Uk_i}\} .$$

Under (1) in tables 2 through 6 we give three different ways of generating p_{Ik_i} based on the congruence $\Sigma\alpha_j x_j \equiv \alpha_0$ (modulo 1) and any integrality restrictions on $x_j, j \in J_N$, using the algorithm in "Cyclic Groups, Cutting Planes, and Shortest Paths". The three different methods simply allow that algorithm to iterate longer. The case of one iteration is the same as Tomlin's work since then the cut is just Gomory's mixed integer cut. The problem in table 4 took longer when the penalties p_{Ik_i} were larger because the trees generated were different. The other problems generally took fewer linear programming iterations and somewhat less time when the penalties were improved. On the pure integer problems, it seems worthwhile to be using these penalties, but for the mixed problems in tables 5 and 6 the simple lp penalty had smaller running times.

Another way of using the mixed integer cut [7] is when the variable x'_j which would be used to satisfy $\Sigma\alpha_j x'_j \geq \alpha_0 - \lfloor\alpha_0\rfloor$, or $\Sigma(-\alpha_j)x'_j \geq \lceil\alpha_0\rceil - \alpha_0$, is an integer variable. When that is the case, that is, when the j giving the minimum in computing p_{Dk_i} is $j = j_0$ where x_{j_0} is an integer variable, then the congruence

$$\sum_{\substack{j \in J_N \\ j \neq j_0}} \frac{\alpha_j}{\alpha_{j_0}} x'_j - \frac{1}{\alpha_{j_0}} s = \frac{\alpha_0 - \lfloor\alpha_0\rfloor}{\alpha_{j_0}}$$

is valid. Furthermore, the cost for satisfying this congruence, using the objective function

$$\sum_{\substack{j \in J_N \\ j \neq j_0}} \left(\gamma_j - \gamma_{j_0} \frac{\alpha_j}{\alpha_{j_0}} \right) x_j' - \gamma_{j_0} \frac{1}{\alpha_{j_0}} s ,$$

can be added onto the lp penalty p_{Dk_i}. In this way p_{Dk_i} can be increased. A similar treatment works for p_{Uk_i}.

In tables 2 through (6), under (2) we give three different ways of using this congruence and the integer restrictions on x_j', $j \in J_N$. Note also that the slack s can also be treated as an integer variable since $s = \lfloor \alpha_0 \rfloor - x_{k_i}$. The three different ways are again simply allowing the cyclic group algorithm to do more iterations.

The mixed problems in tables 5 and 6 were solved in least time using this method and one iteration; that is, the mixed integer cut. For these two problems and for the problem on table 4, doing more iterations did not change the results. For the other two pure integer problems, doing more iterations did in general help although it never paid to do mor than ten iterations.

In conclusion, method (2) using ten iterations seems to be the best of the ones tried. For mixed problems, this number of iterations could be as well set to one. The total times for all five problems were 61, 51, 37, 38, 42, 25, 24, 27 seconds for the eight different methods used (in the order in which they appear in tables 2 through 6). The total lp iterations were 2798, 2551, 1269, 1159, 1269, 1042, 913, 924.

TABLE 1

PROBLEM DATA

Name	Rows	Columns	Integer	0-1	Iter to lp	lp. obj	int. obj
J68	6	8	8	0	9	41.86	39.0
NY2835	28	35	0	35	28	520.0	550.0
NY1244	12	44	0	44	11	56.6	73.0
ANDELU	37	34	0	14	48	1558.3	1492.9
SEVI1	28	44	0	25	64	126,476	119,439

TABLE 2

PROBLEM: J68

Method	Time	Major	no. lp	iter	lift iter
priority	5.2	104	265	513	0
lp pen.	2.4	30	97	209	0
(1) iter = 1	2.7	26	94	214	218
(1) iter = 10	2.1	21	68	139	580
(1) iter = 20	2.5	21	66	135	1160
(2) iter = 1	2.0	18	82	167	384
(2) iter = 10	2.3	17	69	147	1762
(2) iter = 20	5.0	16	71	158	4765

TABLE 3

PROBLEM: NY 2835

Method	Time	Major	no. lp	iter	lift iter
priority	13.8	63	153	736	0
lp pen	14.1	50	158	812	0
(1) iter = 1	9.6	30	85	468	143
(1) iter = 10	9.5	47	93	329	284
(1) iter = 20	9.2	47	93	329	399
(2) iter = 1	7.5	37	87	339	223
(2) iter = 10	5.0	25	55	230	351
(2) iter = 20	5.0	25	55	230	428

TABLE 4

PROBLEM: NY 1244

Method	Time	Major	no. lp	iter	lift iter
priority	9.8	41	101	313	0
lp pen	27.4	107	260	1053	0
(1) iter = 1	6.3	20	51	154	131
(1) iter = 10	9.2	29	70	267	289
(1) iter = 20	13.2	41	99	381	1059
(2) iter = 1	2.7	8	27	91	154
(2) iter = 10	2.9	8	27	91	383
(2) iter = 20	3.3	8	27	91	858

TABLE 5

PROBLEM: Andelu

Method	Time	Major	no. lp	iter	lift inter
priority	12.2	60	186	496	0
lp pen	4.1	18	58	163	0
(1) iter = 1	6.0	19	63	168	215
(1) iter = 10	5.7	19	61	166	487
(1) iter = 20	5.7	19	61	166	496
(2) iter = 1	3.5	15	52	137	102
(2) iter = 10	3.6	15	52	137	220
(2) iter = 20	3.6	15	52	137	225

TABLE 6

PROBLEM: SEVI 1

Method	Time	Major	no. lp	iter	lift iter
priority	20.0	48	387	740	0
lp pen	10.4	28	122	314	0
(1) iter = 1	12.1	28	120	265	241
(1) iter = 10	11.7	28	117	258	550
(1) iter = 20	11.7	28	117	258	551
(2) iter = 1	9.8	27	118	308	330
(2) iter = 10	10.1	27	118	308	909
(2) iter = 20	10.2	27	118	308	910

REFERENCES

1. Beale, E. M. L. and R. E. Small, "Mixed Integer Programming by a Branch and Bound Technique" in: Proc. IFIP Congress 1965, Vol. 2, ed. W. A. Kalenich, Spartan Press, pp. 450-451.

2. Benichou, M., J. M. Gauthier, P. Girodet, G. Hentges, G. Ribiere and O. Vincent, "Experiments in Mixed-Integer Linear Programming", Mathematical Programming 1 (1971) 76-94.

3. Dakin, R. J., "A Tree Search Algorithm for Mixed Inter Programming Problems", The Computer Journal B (1965) 250-255.

4. Driebeek, N. J., "An Algorithm for the Solution of Mixed Integer Programming Problems", Management Science 12 (1966) 576-587.

5. Johnson, E. L. and K. Spielberg, "Inequalities in Branch and Bound Programming", IBM RC 3649, Yorktown Heights, N. Y.

6. Land, A. H. and A. G. Doig, "An Automatic Method of Solving Discrete Programming Problems", Econometrica (1960) 497-520.

7. Tomlin, J. A., "Branch and Bound Methods for Integer and Non-Convex Programming", in: Integer and Non-linear Programming (1970), ed. J. Abadie, American Elsevier Publishing Company, New York, pp. 437-450.

Mathematical Sciences Department
IBM Watson Research Center
Yorktown Heights, New York. 10598

Simplicial Approximation of an Equilibrium Point for Non-Cooperative N-Person Games

C. B. GARCIA, C. E. LEMKE, AND H. LUETHI

1. Introduction.

In this paper we present a method, based upon the 'adjacent simplex' technique discovered and initially developed by Scarf (see, e.g., Ref. [3]), for approximating an equilibrium point of a non-cooperative N-person game. The method also takes advantage of the observations of Wilson (Ref. [10]) and Rosenmüller (Ref. [8]), and their use of 'almost-complementary path'; coupling this with Scarf's generation, in a triangulated simplex, of an 'almost-completely-labelled simplex' path of simplices. In fact, one may consider the method as a way of rendering the (partially constructive) formulation of Wilson-Rosenmüller computable. It may also be considered a generalization of Scarf's initial result in the sense that, for $N = 1$, the method gives a variation of Scarf's approximation for fixed-points.

To test the method, a FORTRAN program was written and equilibrium points for some games, with $N = 3$ and 4, were approximated.

As a brief historical development: in 1963 a constructive proof was given that an equilibrium point exists for a bimatrix game (Ref. [6]). This, in turn, suggested the constructive, combinatorial approach of Scarf for finding, among other things, approximations to fixed-points of continuous mappings.

Scarf's technique was subsequently generalized, refined, and made more amenable for computations by Scarf and Hansen, Kuhn, and Eaves (see Refs. [3], [5], and [2]). Concurrent with this development, Wilson and Rosenmüller (Refs. [10], and [8]), and, differently, Sobel (Ref. [9]) generalized to N-persons the result in [6], using the theme of 'almost complementary path'. The non-constructive character of the Wilson-Rosenmüller result is due to the (relatively nice) non-linearity of the loss functions. By contrast, a truly constructive realization of their observation is given by Howson (Ref. [4]) for the simple polymatrix N-person game. The essence of their approach is summarized in Ref. [7].

The gradual development of the method is the subject matter of Section 2. In 2.1 the proper formulation of the game is given; 2.2 reviews the essentials of a 'general' triangulation of the underlying simplex, embodying some new features; in 2.3 the notion of p-labelling of points and simplices is discussed; in 2.4 the essence of the method is identified; namely the identification of possible 'states' in the labelled triangulation, and of the 'permissible changes of state' (at most two) which permits, in 2.5 the 'standard' almost-complementary path argument put in a simple Lemma in graph-theoretic form.

In Section 3, the triangulation is restricted to the 'standard' one, apparently most useful for computations.

In Section 4, some remarks are made relating to the method, which includes some observations and some suggestions for additional work.

Finally, the results of the computations that were made are summarized in Section 5.

2. Development of the Method.

2.1. The Equilibrium-Point Problem and the Approximation.

For the non-cooperative N-person game, player n has m_n pure strategies; $n = 1, 2, \ldots, N$. The index (n, i) will be used to refer to the ith pure strategy of player n. We write: $m = m_1 + m_2 + \ldots + m_N$; and $\bar{m} = m_1 \cdot m_2 \cdot \ldots m_N$.

The given real quantities:

(1) $a(n; i_1, i_2, \ldots, i_N)$; for $i_n = 1, 2, \ldots, m_n$; $n = 1, 2, \ldots, N$

give the loss per play to player n if player j plays his i_jth pure strategy; $j = 1, 2, \ldots, N$. These quantities define the game, and are considered positive without loss of generality.

Proceeding to mixed strategies: let $x(n, i)$ denote the relative frequency with which n plays his ith pure strategy. Player n's mixed strategy (probability) vector is:

(2) $$x(n) = \langle x(n,1), x(n,2), \ldots, x(n, m_n)\rangle ;$$

of order m_n by 1. Thus, $x(n) \geq 0$, and $e'x(n) = 1$; where $e = \langle 1, 1, \ldots, 1\rangle$. The vector x of variables of the problem:

(3) $$x = \langle x(1), x(2), \ldots, x(N)\rangle ;$$

is thus m by 1. (We use the 'functional' notation, such as $x(n, i)$, rather than $x_{n,i}$, when it avoids awkwardness.)

The function $f(n, i; x)$ defined by:

(4) $$f(n, i; x) = \sum a(n; i_1, \ldots, i_{n-1}, i, i_{n+1}, \ldots, i_N) \cdot \prod_{k \neq n}^{N} x(k, i_k)$$

is the (average) marginal loss to player n if player j plays according to $x(j)$ for $j \neq n$, and player n always plays his ith pure strategy; where the sum in (4) has $\bar{m}/m_n$ terms; one for each combination of pure strategies for the other N-1 players.

Thus we note that

(5.1) $f(n, i; x)$ is independent of $x(n)$;

(5.2) for $j \neq n$, $f(n, i; x)$ is homogeneous of degree 1 in $x(j)$; thus $f(n, i; x)$ is homogeneous of degree N-1 in x; that is: $f(n, i; tx) = t^{N-1} f(n, i; x)$;

(5.3) $f(n,i;x) > 0$ if and only if, for all $j \neq n$: $x(j) \neq 0$; hence, for $j_1 \neq j_2$: $f(j_1, i;x) = 0$ if $x(j_2) = 0$.

The function $g(n;x)$ defined by:

$$g(n;x) = \sum_{i=1}^{m_n} x(n,i) \cdot f(n,i;x) \tag{6}$$

is the average (long-run) loss per play of player n if plays are made according to the strategy vector x.

A (Nash) equilibrium point for the game is a strategy vector x satisfying:

$$g(n;x) \leq f(n,i;x); \quad \text{for all } i = 1,2,\ldots,m_n; \; n = 1,2,\ldots,N. \tag{7}$$

These conditions for an equilibrium point are equivalent to the (complementary) conditions:

$$\begin{aligned} &f(n,i;x) \geq v(n); \; x(n,i) \cdot [f(n,i;x) - v(n)] = 0; \\ &x(n) \geq 0; \; v(n) \geq 0; \text{ and } e'x(n) = 1; \text{ for all} \\ &i = 1,2,\ldots,m_n; \; n = 1,2,\ldots,N. \end{aligned} \tag{8}$$

Given x, define $k(x)$ by:

$$k(x) = \operatorname{Min.}_{(n,i)} f(n,i;x). \tag{9}$$

The homogeneity of the f's permits the equivalence: (where the requirement: $e'x(n) = 1$ is dropped)

$$\begin{aligned} &x(n,i) \cdot [f(n,i;x) - k(x)] = 0; \\ &x \geq 0 \text{ and } x(n) \neq 0; \; n = 1,2,\ldots,N \end{aligned} \tag{10}$$

(that is: either $x(n,i) = 0$ or $f(n,i;x) = k(x)$).

Actually, this complementary problem is the one for which the method to be described yields an approximate solution. In the process, no assumptions concerning 'degeneracy' need be made. We identify the approximation as follows:

> For a point x we shall say that an index (n, i) is <u>complementary</u> at x if and only if (10) is satisfied for (n, i).

When the method terminates, we shall have a simplex, call it $[s]$; of dimension $M-1$; $N \leq M \leq m$; of small diameter; with some M vertices $x_1, x_2, \ldots, x_M$; and some M distinct indices: $(r_1, s_1), (r_2, s_2), \ldots, (r_M, s_M)$ such that:

(11.1) each (n, i) which is not some (r_t, s_t) is complementary at each vertex of $[s]$ (in fact: $x(n, i) = 0$ on $[s]$); and for each (n, i) which is some (r_k, s_k); $x(n, i)$ is positive on the interior of $[s]$;

(11.2) (r_t, s_t) is complementary at x_t; $t = 1, 2, \ldots, M$;

(11.3) $x_t(n) \neq 0$; $n = 1, 2, \ldots, N$ and $t = 1, 2, \ldots, M$.

Having such $[s]$, to revert to probabilities, set:

$$\bar{x}(n) = x(n)/e'x(n) .$$

We may use any point of $[s]$ to approximate: let us focus upon x_1. Write: $P = [e'x_1(1)] \cdot [e'x_1(2)] \cdot \ldots [e'x_1(N)]$.

By homogeneity we have:

$$f(r_t, s_t; x_1) = [P/e'x_1(r_t)] \cdot f(r_t, s_t; \bar{x}_1) . \tag{12}$$

If we write: $v(t) = k(x_t)e'x_1(r_t)/P$, we may evaluate the 'deviation from complementarity' (8) at $\bar{x}_1$ as:

$$\sum_{t=2}^{M} \bar{x}_1(r_t, s_t) \cdot [f(r_t, s_t; \bar{x}_1) - v(t)] ; \tag{13}$$

which, since the f's are continuous, and $[s]$ has small diameter, approximates 0.

2.2 Triangulation of the Simplex

We shall triangulate the simplex $S(K)$, where:

(14) $$S(K) = \{x \geq 0 : e'x = K\} ;$$

where K is a fixed positive number, and (for our purposes) $K > N$. We shall write S for $S(K)$. (See, for example, Ref. [1] for definitions of triangulation.)

We are given a set of points of S (the Mesh), which includes the vertices of S, which comprises the set of vertices of simplices in the triangulation of S. We denote by $[s]$ a typical member of triangulation.

A boundary-face of S may be defined by some M of the m vertices of S. The boundary-face, call it S', thus identifies some M indices (n, i) such that $x(n, i)$ is positive on the interior of S'. Thus, S' has dimension $M-1$ (and $S' = S$ when $M = m$). The triangulation triangulates S and each of its boundary-faces.

Each boundary-face of S of dimension $M-1$ is triangulated into simplices $[s]$ of dimension $M-1$ (with M vertices) in such a way that a simplex $[s]$ in the boundary-face is a member of the triangulation iff it is a face of some simplex in the triangulation of dimension M. (Actually, in the method, only simplices $[s]$, for which $M \geq N$ are encountered.)

Adjacent Simplices.

We next identify those simplices $[s]'$ of the triangulation which are adjacent to a given $[s]$. Adjacency is defined by the following rules (used throughout) for forming another $[s]'$ from $[s]$.

(15.1) Drop Vertex. Consider a vertex, call it V_d, of $[s]$. (Note: for a current simplex $[s]$, to avoid confusion, we label vertices with 'V', rather than 'x'.) Let F denote the convex hull of the remaining $M-1$ vertices of $[s]$ (F is the 'face opposite' V_d). Intrinsic to the triangulation of S, there is a unique member [s]' of the triangulation, obtained from [s] by dropping V_d and either:

(15.1.1) adjoining a new vertex, call it V_a, (which thus forms [s]', having the same dimension, M - 1, as [s], and lies in the same bounding-face of S);

or

(15.1.2) not adjoining a new vertex (in which case [s]' = F, of dimension M - 2. Thus, this is the case where, for some (n, i) we have that x(n, i) is positive on [s], but is 0 on [s]' = F.)

(Note: for some [s] in the triangulation, if for some (n, i) x(n, i) is positive at some vertex of [s] -- and this holds iff x(n, i) is positive in the interior of [s] -- we say that 'x(n, i) is positive on [s]'.)

(15.2) Add vertex. (The 'inverse' of (15.1.2)). Let (n, i) be an index such that x(n, i) is 0 on [s]. Then there is a unique Mesh point, call it V_a, for which $V_a(n, i)$ is positive, such that the simplex [s]', with vertices V_a and all vertices of [s], is a member of the triangulation.

For the triangulation, these operations are 'reversible' in the sense that, (i) for case (15.1.1) "drop V_d, add V_a", doing (15.1) on [s]' with V_a to be dropped would be (15.1.1), with V_d the added vertex; and (ii), as noted, operations (15.1.2) and (15.2) are 'inverses'.

(We simply state without proof that, from any simplex in the triangulation, one may go to any other by a finite number of such operations; that S is the union over the triangulation; and that two members of the triangulation having the same dimension intersect in a set at most one dimension less.)

2.3 The Labelling Process.

Consider the sets of indices:

(22) $I(0) = \{(1,1), (2,1), \ldots, (N,1)\}$;

$B(p) = \{(p,2), (p,3), \ldots, (p,m_p)\}$;

and

$$I(p) = I(p-1) \cup B(p)$$

$$= I(0) \cup B(1) \cup B(2) \cup \ldots \cup B(p) ;$$

$p = 1, 2, \ldots, N$.

Given a point x in S , we say that (n,i) is the p - label for x if and only if:

(22.1) (n,i) is in $I(p)$; and

(22.2) (n,i) is the least* index in $I(p)$ such that:

$$f(n,i;x) = \operatorname*{Min.}_{(r,s) \in I(p)} f(r,s;x) .$$

*'least' refers to the ordering of the (n,i), which is lexicographic: e.g., $(1,4)$ is less than $(2,1)$.

We may note that:

(22.3) If for some q , $(q,1)$ is the q -label for some point x , then $(q,1)$ is the q-1 -label for x .

Let $[s]$ be a member of the triangulation with $M \geq N$ vertices; hence some M components $x(n,i)$ are positive on $[s]$. By the p-labelling of $[s]$ we mean the p-labelling of its M vertices. By (22), for each p , $[s]$ has a unique p-labelling.

An index (n,i) is p-complementary at a point x of S if and only if either $x(n,i) = 0$, or (n,i) is the p-label of x .

p -proper Labelling of [s].

We shall say that the p-labelling of [s] is p-almost complementary if and only if for all indices (n, i) except (p, 1), (n,i) is p-complementary. Hence: x(p, 1) is positive on [s] and (p, 1) is not a p-label.

Similarly, we shall say that the p-labelling of [s] is (p+1)-almost complementary if and only if for all indices (n, i) except (p+1, 1), (n, i) is p-complementary. Hence: x(p+1, 1) is positive on [s] and (p+1, 1) is not a p-label for [s].

In particular, the p-labelling of [s] is not both p-almost complementary (p-AC) and (p+1)-almost complementary ((p+1)-AC).

We shall say that [s] is p-complementary if and only if each index (n, i) is p-complementary.

We shall say that [s] is p-proper if and only if:

(23.1) [s] is either p-AC, (p+1)-AC, or p-complementary,

(23.2) for all q: for some i, x(q, i) is positive on [s].

In particular, these imply:

(23.3) (n, i) is p-complementary on [s] for all i, and

$$n = 1, 2, \ldots, p-1\,;$$

for $n = p+1, \ldots, N$: x(n, 1) is positive on [s], and

$$x(n, i) \text{ is } 0 \text{ on } [s] \text{ for } i \neq 1\,.$$

2.4 Possible States and Permissible Changes of State.

Associated with each iteration of the method is a simplex [s] of the triangulation and a specified p such that [s] is p-proper. We shall write this as (p; [s]).

Note that, if [s] is p-proper, it may also be (p+1)-proper, or (p-1)-proper; more generally, may also be q-proper for certain $q \neq p$. However, for the method, we shall want to consider, in particular, for the same [s], (p; [s]), and (for example) (p+1, [s]) as <u>different</u> states.

In more detail, we shall:

<u>i.</u> for a fixed (generic) pair (p; [s]), delineate all possible states;

<u>ii.</u> identify <u>permissible</u> changes of state from each possible state, for (p; [s]). In fact, these changes of state will either be to a state for (p; [s]'); (where [s]' is a uniquely determined simplex adjacent to [s]); or to a state for (q; [s]) (no change in [s]) for q = p-1 or p+1 .

This will be done in such a way that:

<u>iii.</u> for each state of (p; [s]), one or two changes of state will be permissible. In fact, there will be exactly <u>one</u> for a unique <u>starting</u> simplex, $[s]_0$ (to be identified) or for a state (N; $[s]_T$) for which $[s]_T$ is N-complementary (a 'terminal' state). For all other states, there will be exactly two changes of state permitted.

Those changes of state from a (p; [s]) state to a (p; [s]') state will be effected by a <u>Drop</u> <u>Vertex</u> or <u>Add Vertex</u> ((15.1.1), (15.1.2), or (15.2)).

We shall always assume that we are not in the initial state, nor in a terminal state, so that there will be exactly two changes of state to identify (uniquely).

We shall first accumulate some observations.

<u>[s] is p-proper.</u>

First, suppose that [s] is p-AC or (p+1)-AC . Then, since [s] has some $M \geq N$ positive components, for <u>exactly</u> M-1 of the M (n, i) such that x(n, i) is positive on [s], (n, i) is a p-label (and if p-AC, (p, 1) is <u>not</u> a p-label; if (p+1)-AC, (p+1, 1) is not a p-label).

In either case, there will be a unique (r, s) in I(p) such that either:

(24.1) (r, s) is the p-label for exactly two vertices, call them V_1 and V_2 (hence x(r, s) is positive on [s]);

or

(24.2) (r, s) is the p-label for one vertex of [s], call it V_0, and x(r, s) = 0 .

(Note: most of the Drop- or Add-Vertex operations will be dropping V_0, V_1, or V_2, and adding a vertex making x(r, s) positive.)

Next, suppose that [s] is p-complementary. Then, either [s] is as in (25.1), or is not, and either [s] is (p-1)-proper or is not. We examine each of these possibilities:

(25.1) the p-labelling of [s] is (p+1)-proper, but its (p+1)-labelling is not (p+2)-AC .

Since [s] is p-complementary, x(p+1, 1) is positive on [s], and (p+1, 1) is a p-label, for some vertex V of [s].

Then, a fortiori, the (p+1)-label of V is, in any case, (p+1, i) for some i .

Now, for V' ≠ V , if (r, s) is the p-label for V' , x(r, s) is positive on [s] and (r, s) ≠ (p+1, 1); since [s] is p-complementary. Hence, (r, s) is also the (p+1)-label for V' , since [s] is as in (25.1).

Hence, the (p+1)-labelling of [s] is (p+1)-AC if and only if the (p+1)-label for V is (p+1, i) for i ≠ 1 ; whereas [s] is (p+1)-complementary if and only if (p+1, 1) is also the (p+1)-label for V .

(25.2) [s] is (p-1)-proper.

If x(q, 1) is positive on [s], (for either q = p-1, or q = p) then, since [s] is p-complementary, (q ,1) is a p-label for some V ; hence is also a (p-1)-label for V . Hence the (p-1)-labelling of [s] is neither (p-1)-AC nor p-AC. Hence [s] is <u>(p-1)-complementary.</u>

Hence:

<u>i</u> x(p,i) is positive on [s] if and only if i = 1

<u>ii</u> (p,i) is a p-label if and only if i = 1 .

(Note: Hence, if for some q , [s] is both q-complementary, and (q+1)-complementary, for each V , the q- and the (q+1)- labels are the same -- the q-labelling is the (q+1)-labelling.)

(25.3) [s] is not as in (25.1).

Again, let V denote the unique vertex with p-label (p+1,1), (hence, a fortiori, (p+1,k) is the (p+1)-label for V, for some k).

Since [s] is not as in (25.1), there is some vertex V' ≠ V such that the (p+1)-label for V' is (p+1,i) for some i . Hence, <u>at least two</u> vertices of [s] have labels of the form (p+1,i).

(Note: In the case (25.3), therefore, if one were to <u>Drop V</u>, for the resulting [s]', at least one vertex will have the (p+1)-label (p+1,i) for some i . We shall use this result.)

(25.4) [s] is not (p-1)-proper (complementary).

Since [s] is p-complementary, either x(p,1) = 0 or (p,1) is a p-label for some V , (hence, the (p-1)-label for V).

Since [s] is not (p-1)-complementary, for some V' ≠ V, (p,i) is the p-label for V', for some i , where i≠1.

(Note: in this case, we will want to consider <u>Add</u> <u>Vertex</u>, making x(p,1) positive, in case x(p,1) = 0 ; and will want to <u>Drop V</u> in case (p,1) is a p-label. In the latter case, then, for the resulting [s]', (p,i) will be a p-label for some i . We shall use this result.)

We next identify the possible states for the case (p; [s]). Concurrently, for each delineated state, we identify the two permissible changes of state.

The Figure shows, in brief, the possible states and the permissible changes of state.

Note that branches labelled "n" indicate no change in [s] in the process of changing states. All other branches involve a change in [s].

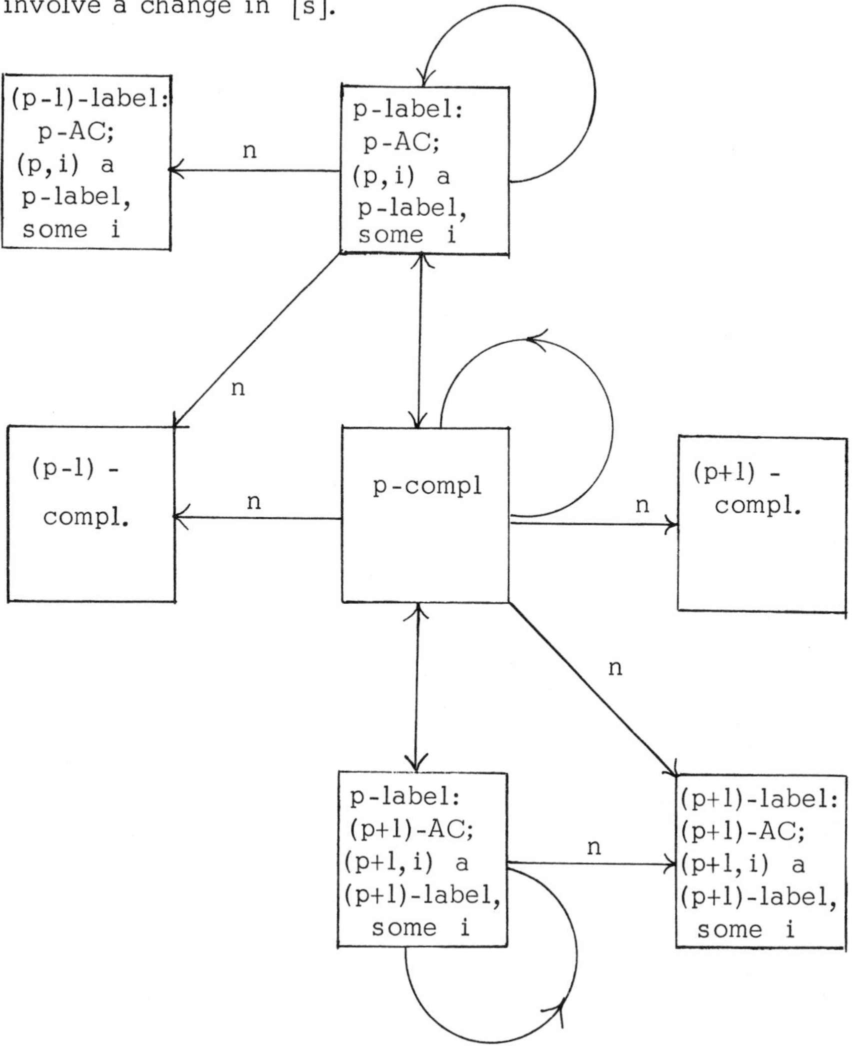

Case 1. [s] is p-AC and (p,i) is a p-label for some i .
Let (r,s) be as in (24).

Case 1.1 (24.1).

State Changes: 1. Drop V_1 .

2. Drop V_2 .

(Note: The actual resulting state is implied. Thus, the reader may ascertain that i. the resulting [s]' is again p-proper, and that ii. that state, which cannot be ascertained until the operation is effected, is again one of the cases we are in the process of delineating, for (p; [s];).)

Case 1.2 (24.2).

Case 1.2.1. There is an i such that (r, i) is a p-label and $i \neq s$.

State Changes: 1. Drop V_0 .

2. Add V_a (by increasing x(r, s) from 0).

Case 1.2.2. (r, i) is a p-label only for i = s . Hence r = p; $s \neq 1$.

(Note: Since x(r, s) = 0 on [s], x(r, i) is positive on [s] for some $i \neq s$; hence, since [s] is p-AC and (r, i) is not a p-label for $i \neq s$, (r, i) = (p, 1); hence $s \neq 1$.)

State Changes: 1. Some (unique) state for (p-1, [s]).

2. Add V_a (by increasing x(r, s) from 0).

(Note: The condition for Case 1 that (p, i) is a p-label for some i is imposed in order that permissible changes result in possible states. Case 1.2.2. is delineated from Case 1.2.1 to insure that the Case 1 condition noted holds. Thus, in Case 1.2.2, one has (p, i) a p-label only for i = s , so that were one to Drop V_0, one would run the risk of violating the Case 1 condition.)
(Note: Again, by State Change 1 , we are implying that the actual state is some state we are delineating, with p replaced by p-1. Actually, since [s] does not change, one

could ascertain the state, but this would be redundant. However, in this case, it is readily deduced that the state is either a Case 2 or Case 3 state.)

Case 2. (p+1)-AC and (p+1, i) is a (p+1)-label for some i.

Case 2.1. [s] is (p+1)-proper.

Let V_1 be a vertex with (p+1)-label (p+1, i) for some i. Then since, by (p+1)-AC, (p+1, 1) is not a p-label, it is not a (p+1)-label; hence $i \neq 1$. Moreover, since x(p+1, 1) is positive on [s], the (p+1)-labelling of [s] is (p+1)-AC.

Hence, for all M-1 (n, i) such that x(n, i) is positive on [s], and $(n, i) \neq (p+1, 1)$, (n, i) must be a (p+1)-label for some vertex of [s]; which is also its p-label, since, by p-proper, $n \neq p+1$. Let (r, s) be the p-label of V_1. Then, $r \neq p+1$, by (p+1)-AC. Now, if (r, s) is the p-label for V_1 only, then x(r, s) = 0 (since x(r, s) positive would imply that (r, s) is also the (p+1)-label, for V' ($\neq V_1$), say; and hence, (r, s) is also the p-label of V', a contradiction). Hence, (r, s) is that of (24).

Case 2.1.1. (r, s) p-labels V_1 and V_2.

State Changes: 1. Some unique state for (p+1; [s])

2. Drop V_2

(Note: Again, Drop V_1 is not permissible in this case, since we run the risk of violating the Case 2 condition; which in turn is the Case 1 condition for some (p+1, [s]')).

Case 2.1.2 (r, s) p-labels only V_1 (hence, x(r, s) = 0).

State Changes: 1. Some unique state for (p+1; [s])

2. Add V_a (increasing x(r, s) from 0).

(Note: Of course, the unique states for (p+1, [s]) are for (p+1)-AC, since [s] does not change, and the condition for Case 1 is thus satisfied.)

Case 2.2 [s] is not (p+1)-proper. Then, for (r, s) as in (24):

Case 2.2.1. (r, s) p-labels V_1 and V_2

State Changes: 1. Drop V_1

2. Drop V_2.

Case 2.2.2 (r, s) p-labels V_0 and $x(r,s) = 0$

State Changes: 1. Drop V_0

2. Add V_a (increasing $x(r,s)$ from 0).

(Note: Compare with Case 1.2. Case 1.2.2. is inapplicable here since, $r \neq p+1$ implies there is some $x(r,i)$ positive on [s]; hence (r, i) is a p-label, and $(r,i) \neq (r,s)$.)

Case 3. [s] is p-complementary.

We have delineated the four possibilities in (25); which in turn give rise to four cases:

Case 3.1 [s] is (p-1)-proper and is of the form (25.1).

State Changes: 1. Some unique state in (p+1, [s])

2. Some unique state in (p-1, [s])

Case 3.2 [s] is not (p-1)-proper but is of the form (25.1).

Case 3.2.1 (p, 1) is a p-label, for V, say.

State Changes: 1. Some unique state in (p+1;[s]).

2. Drop V.

Case 3.2.2 $x(p,1) = 0$ on [s].

State Changes: 1. Some unique state in (p+1, [s]).

2. Add V_a (increasing $x(p,1)$ from 0).

Case 3.3. [s] is (p-1)-proper and not of the form (25.1).

State Changes: 1. Drop V (in (25.3)).

2. Some unique state in (p-1, [s]).

Finally, for completeness; with obvious state changes:

Case 3.4. [s] is not (p-1)-proper nor is of the form (25.1).

Case 3.4.1 (p, 1) is a p-label for V.

Case 3.4.2 $x(p, 1) = 0$.

We terminate this sub-section with an observation which is an essential feature of our approach; namely that, for each simplex [s] arrived at: $x(n, i)$ is positive on [s] for some i, for each $n = 1, 2, \ldots, N$. (Note that, in the operation: Drop vertex, if there is no added vertex, so that some $x(r, s)$ becomes 0 on [s], conceivably on the new simplex [s]', one could have $x(r, i) = 0$ on [s]', for all $i = 1, 2, \ldots, m_r$. We point out that this cannot happen.

Lemma: Let [s] have $M \geq N$ vertices. Suppose that [s] is p-proper. Then, at the next simplex [s]' generated by the method, for all j, there is an i such that $x(j, i)$ is positive on [s]'.

Proof: Let us prove the lemma for $N \geq 3$ (the proof is similar for $N = 2$, and trivial for $N = 1$).

Suppose that, at a vertex V of [s], for some r, $x(r) = 0$. Then $f(n, i; V) = 0$ for all $i = 1, 2, \ldots, m_n$ and $n \neq r$. Hence, the p-label for V, for all p, is either (1, 1) or (2, 1).

Now, suppose for contradiction, that $x(r) = 0$ on [s]'. Then since $x(r) \neq 0$ on [s], we must have dropped a vertex V_d (in which case, [s'] is a face of [s]). Hence, the p-labels at [s]' are all different, and are either (1, 1) or (2, 1). Thus, $M = N = 3$.

Now, note that the p-label at V_d cannot be (3, i), else, by Case 1.2.2, we cannot drop V_d. Hence, we must be 3-AC with respect to I(2), since (3, i) is not a p-label. By Case 2, (3, i) must be the 3-label at V_d, the 3-labels at [s]' being either (1,1) or (2,1). Hence, we are 3-proper, and by Case 2.1, we cannot drop V_d. This completes the proof for $N \geq 3$.

2.5. The Almost-Complementary Path. A Lemma on Graphs.

The method being proposed may be viewed as one generating a sequence of 'adjacent' states, starting from a uniquely-defined starting state, proceeding in a unique manner until a terminal state is arrived at, which corresponds to an N-complementary simplex.

In more detail, we first define a starting simplex $[s]_0$, based upon I(0), and a starting state for the pair (1, $[s]_0$) such that there is just one permissible Change of State.

Having such a START, we have the familiar argument, (which, however, we substantiate later as an application of a lemma involving graphs):

If, in the process of going from state to state by permissible changes, one arrives at a state which is not a terminal state, one does so by one of the exactly two permissible changes of state. Since one may then make the unique other permissible change of state, the iterations do not then terminate and, when leaving a particular state, one cannot return to it, so that no state repeats (see the following lemma). Since the number of pairs (p, [s]) is finite one must terminate; hence in a terminal state, from which the desired approximation to an equilibrium point may be made.

The Starting Simplex $[s]_0$.

Let e(n, i) denote the m by 1 column with 1 as (n, i)th component, and 0 as other components.

Consider the bounding face, S_0, of S which is the convex hull of the N points: Ke(n, 1); n = 1, 2, . . . , N. We shall assume that we have a simplex $[s]_0$ in the relative interior of S_0 (i.e., x(n, i) is positive on $[s]_0$ for i = 1, and is 0 on $[s]_0$ for $i \neq 1$; $n = 1, 2, \ldots, N$).

We assume that $[s]_0$ is 1-proper but not 2-AC at I(1), <u>and</u> (the condition for State 1) that, for some q, $(1, q)$ is a 1-label for $[s]_0$. We assume that $[s]_0$ is a member of the triangulation and is the only member with the above properties.

Note that, since $[s]_0$ is 1-proper, $(n, 1)$ is a 1-label for $[s]_0$, for $n = 2, 3, \ldots, N$, and that $(1, q)$ is the only label of the form $(1, i)$. The starting state is one of the states for $(1, [s]_0)$: it is of the <u>State 3</u> (1-complementary) if and only if $q = 1$; of the <u>State 1</u> if and only if $q \neq 1$.

A Lemma on Graphs.

Let $G = G(T, B)$ denote a graph, with t in T a node of G; b in B a branch (unordered) of B.

We suppose that each b in B connects two distinct nodes; and that two distinct nodes are unconnected or connected by just one branch.

Let $B(t)$ denote the set of branches incident on node t. Without loss of generality, we take $B(t) \neq \phi$.

Let N denote a fixed positive integer.

Consider a function g defined on B by:

$$g(b) = (r_1, r_2, \ldots, r_N);$$

where r_i is a non-negative integer.

g induces a $\bar{g}$ on T by:

$$\bar{g}(t) = \sum_{b \text{ in } B(t)} g(b) .$$

Such a g we call <u>permissible</u> if it has the following properties:

For each t in T ; for each p ; $1 \leq p \leq N$:

either (1) $\bar{g}_p(t) = 0$ (in which case we say that t has 'power 0 at level p'),

or (2) $\bar{g}_p(t) \geq 1$, in which case we define the 'power of t at level p' as:

$$\bar{g}_{p-1}(t) + \bar{g}_p(t) + \bar{g}_{p+1}(t)$$

with the understanding that, cyclically (or modulo N), if $p = 1$; $p-1$ is replaced by N ; if $p = N$, $p+1$ is replaced by 1 ; and:

(2a) for $p = 2, 3, \ldots, N-1$, the power of t at level p is 2 ;

(2b) for $p = 1$ or $p = N$ the power of t at level p is 1 or 2 .

(In particular, $g_i(b)$ is 0, 1, or 2 for permissible g.)

Note that then the power of t cannot be 1 at both level 1 and level N . Hence we will say that the 'power of t is 1', if either at level 1 or level N .

Lemma: If g is permissible, an even number of nodes have power 1.

Prior to giving the proof, let us note the application of the Lemma to the method of the text. A node t in T corresponds to a state (p; [s]); and adjacent states are such that their corresponding nodes are connected by a branch. The Lemma is proved 'constructively' by pairing-up uniquely pairs of nodes having power 1 by means of a 'g-path'.

For the method of the text, the graph G has just one node with power 1 at level 1 ; namely $(1, [s]_0)$; and terminal states (N, [s]) correspond to nodes of power 1 at level N .

Proof of Lemma. If there are no nodes of Power 1 , since 0 is even the Lemma holds. Otherwise, let t_0 be a node of power 1 . We shall uniquely generate a path -- which we call a 'g-path' -- of nodes:

$$t_0, t_1, \ldots, t_L$$

such that t_L is of power 1 ; $t_L \neq t_0$; and, for $r = 1, 2, \ldots, L$

t_{r-1} and t_r are connected by a branch, which we call b_r. The g-path from t_0 is generated as follows:

We have either $\bar{g}_1(t_0) = 1$; or $\bar{g}_N(t_0) = 1$, and not both. By the symmetry of g we may suppose, without loss of generality, that $\bar{g}_1(t_0) = 1$. Hence there is one and only one b in $B(t_0)$; call it b_1; such that $g_1(b_1) = 1$ (and hence $g_N(b_1) = g_2(b_1) = 0$; and $g_q(b) = 0$, for $q = N, 1$, or 2 for b in $B(t_0)$; $b \neq b_1$).

We associate with b_1 the <u>level</u> $p = 1$.

Generally, suppose that in the generation we have arrived at some node t_r. If $r > 0$, we will have just traversed some branch b_r incident with and leading to t_r, such that some level p has been associated with b_r, and $g_p(b_r) > 0$.

Hence $\bar{g}_p(t_r) = 1$ or 2.

<u>Case 1.</u> $\bar{g}_p(t_r) = 1$. Hence, $p = 1$ or $p = N$, and the g-path is terminated (because, for all b in $G(t_r)$ there is only one $p-1$, p, or $p+1$ component which is non-zero; namely $g_p(b_r) = 1$ -- the subscripts read modulo N).

<u>Case 2.</u> $\bar{g}_p(t_r) = 2$. There is then some <u>one</u> b in $G(t_r)$; call it b_{r+1}, such that:

<u>i.</u> $b_{r+1} \neq b_r$;

<u>ii.</u> $g_q(b_{r+1}) > 0$ for exactly one of $q = p-1,\ p,\ p+1$,

and we traverse b_{r+1} (i.e., identify its other end-point t_{r+1}) and associate with b_{r+1} the level q.

(Note: the definition of permissible g permits that, for some b and p: $g_p(b) = 2$. However, no such branch is traversed in generating a g-path, since then, if for some r $g_p(b_r) = 2$, where b_r was given the associated level p, this would imply that: for $r = 1$, the power of t_0 is not 1; for $r > 1$, the power of t_{r-1} is greater than 2 for level p.)

This completes the description of the generation of the g-path. Note that the p-values associated with branches adjacent in the g-path differ by at most 1.

Continuing the proof, we observe that when we have once traversed some branch b_r at level p we cannot retraverse it at the same level (although traversal of b_r at a different level is generally possible). Thus, suppose, for contradiction, that b_r were retraversed at level p . We may suppose that this is the first such retraversal. Schematically we visualize:

```
               (p)
t    o---------o t  .
 r-1     b        r
          r
```

Thus, we traversed b_r from left to right the first time. Suppose, for example, that a second traversal was from right to left (a similar argument resolves the other case). Thus, schematically:

```
                         (q )
                           s
                    o--------o t  (s ≠ r+1)
    <----------    /           s
         (p)      /       b
t    o-------o           s
 r-1            /   t
        b      /     r
         r    /      (q   )
             /          r+1
            o------------o t
                            r+1
                 b
                  r+1
```

Then both q_s is p-1, p, or p+1; and q_{r+1} is p-1, p, or p+1; so that $g_{q_s}(b_s) > 0$; and $g_{q_{r+1}}(b_{r+1}) > 0$. But then the power of t_r at level p is at least 3 ; a contradiction.

Since the number of combinations of branches and levels is finite, the path must terminate at a node, t_L , of power 1 , and $L \neq 0$. Since t_L is arrived at by traversing at level $p = 1$ or $p = N$ a branch b_{L-1} ; $t_L \neq t_0$ (since $t_L = t_0$ would imply a retraversal of $b_1 = b_{L-1}$ at the same level 1).

Finally, we need to observe the reversibility -- namely that the g-path generated from t_L is the same as the g-path generated from t_0 (but traversed in the opposite direction). This follows from the symmetrical definition of permissible g . This completes the proof.

Above all, we again observe that branches may be retraversed, but always at different levels. For example;

N = 10 (where we have written, for example, 2,6,9 for the 10-tuple (0,1,0,0,0,1,0,0,1,0); which has 1's in the 2, 6, and 9 components):

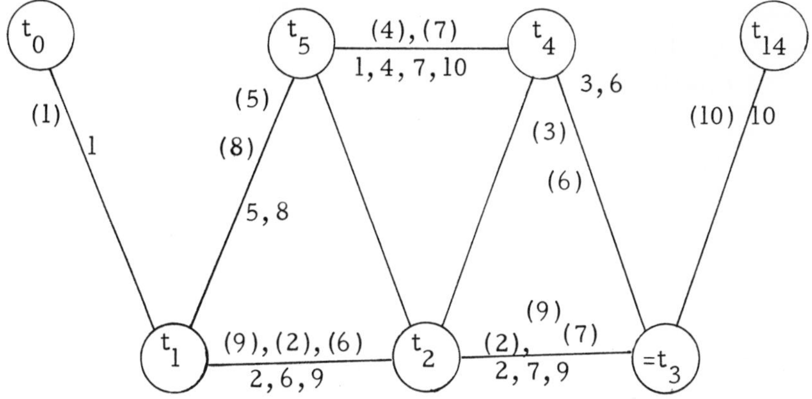

The g-sequence is: $t_0, t_1, t_2, t_3, t_4, t_5, t_6 = t_1$, $t_7 = t_2$, $t_8 = t_3$, $t_9 = t_4$, $t_{10} = t_5$, $t_{11} = t_1$, $t_{12} = t_2$, $t_{13} = t_3$, t_{14}.
(Adding 1 to any component of any N-tuple would give a g which is not permissible.)

Of course, for the method of the text, values $\bar{g}_p(t) = 2$, for example, correspond to the two permissible changes of state.

3. The Standard Triangulation of S.

We shall partially describe the triangulation used for the computations; for which it is the simplest, and easiest for 'recording' simplices, and effecting Drop and Add operations.

To simplify the description of the triangulation, we shall first identify $[s]_0$, and in a unique way, given the explicit game, so that it has the properties previously given. It will be evident that it is unique.

We then describe a typical element [s] of the triangulation, and observe how Drop and Add operations are effected.

3.1 Description of $[s]_0$

Given $K > N$, consider the functions $f(n,i;x)$ for points interior to S_0 (thus, $x(1,1) + x(2,1) + \ldots + x(N,1) = K$; and $x(n,1)$ positive for $n = 1,2,\ldots,N$).

For such x the $f(n,i;x)$ have the simple form:

$$f(n, i; x) = \bar{a}(n,i) \cdot \prod_{k \neq n}^{N} x(k,1) ; \tag{26}$$

or, letting $P = \prod_{n=1}^{N} x(n,1)$:

$$x(n,1) \cdot f(n,i;x) = \bar{a}(n,i) \cdot P. \tag{27}$$

Hence, for any (r,s), (t,u) we have:

$$f(r,s;x) \leq f(t,u;x) \quad \text{if and only if} \quad x(t,1)/x(r,1) \leq \bar{a}(t,u)/\bar{a}(r,s) . \tag{28}$$

Let q satisfy:

$$\bar{a}(1,q) = \operatorname*{Min.}_{i} \ \bar{a}(1,i) \tag{29}$$

(and, if $\bar{a}(1,i) = \bar{a}(1,q)$; then $q \leq i$).

Let $\bar{x}_0$ be the point of S_0 satisfying:

$$f(1,q;\bar{x}_0) = f(2,1;\bar{x}_0) = \ldots = f(N,1;\bar{x}_0) . \tag{30}$$

By (28):

$$\bar{x}_0(n,1) = [\bar{a}(n,1)/\bar{a}(1,q)] \cdot \bar{x}_0(1,1) . \tag{31}$$

Note, from (30), that $(1,q)$ is the 1-label for $\bar{x}_0$. Since $\bar{x}_0$ is in S_0, we have:

$$\bar{x}_0(1,1) = k_0 \bar{a}(1,q) \cdot K/[\bar{a}(1,q) + \sum_{n=2}^{N} \bar{a}(n,1)] ; \tag{32}$$

where k_0 is chosen to make $\bar{x}_0$ integral.

We next specify the vertices: V_i of $[s]_0$ as:

(33) $$V_1 = \bar{x}_0 + e(2,1) - e(1,1) ;$$

$$V_2 = V_1 + e(3,1) - e(2,1);$$

$$\vdots \qquad \vdots$$

$$V_N = V_{N-1} + e(1,1) - e(N,1) .$$

(Note, by summing, that $V_N = \bar{x}_0$).

Let us define direction vectors:

(34) $$w[(r,s), (t,u)] = e(t,u) - e(r,s).$$

$[s]_0$ may be compactly represented by its sequence:

(35) $$\bar{x}_0: [(1,1), (2,1)], [(2,1), (3,1)], \ldots, [(N,1), (1,1)].$$

(Note that the directions sum to zero.)

Note that each component of any two vertices of $[s]_0$ differ by at most 1 ; in fact, for each component (n,i) there is an integer $c(n,i)$ such that for each V_k , $V_k(n,i)$ is $c(n,i)$ or $c(n,i) + 1$. Of course, any two adjacent vertices of $[s]_0$ differ in only two components.

The 1-labelling of $[s]_0$

For a point x with positive components $(n,1)$; $n = 1, 2, \ldots, N$, in order that x has the 1-label $(r,1)$ requires, from (28), that:

(36) $$x(r,1)/x(1,1) \geq \bar{a}(r,1)/\bar{a}(1,q) \quad \text{and}$$

$$x(r,1)/x(n,1) \geq \bar{a}(r,1)/\bar{a}(n,1); \text{ for } n = 2,3,\ldots,N.$$

or, using (31), that:

(37) $$x(r,1)/x(n,1) \geq \bar{x}_0(r,1)/\bar{x}_0(n,1); \quad n = 1,2,\ldots,N .$$

Now, consider V_k, for $k < N$. From (33) we have: $V_k(1,1) = \bar{x}_0(1,1) - 1$; and $V_k(k+1,1) = \bar{x}_0(k+1,1) + 1$; and $\bar{x}_0$ and V_k have equal other components.

Hence, for $n \neq 1$ or $k+1$:

(38) $$V_k(k+1,1)/V_k(n,1) = [1+\bar{x}_0(k+1,1)]/\bar{x}_0(n,1) > \bar{x}_0(k+1,1)/\bar{x}_0(n,1)$$

so that (37) is satisfied for such n; we similarly find that (37) is satisfied for $n = 1$ and $k+1$.

Hence, $(k+1,1)$ is the 1-label for V_k; $k = 1,2,\ldots, N-1$.

It follows that $[s]_0$ has the desired properties.

3.2 A Simplex of the Triangulation.

For the given K, the Mesh of the triangulation consists of all the integral points of S.

A simplex $[s]$ is a member of the triangulation if and only if it may be represented by a sequence:

(39) $$\bar{x};\ [(r_1,s_1),\ (t_1,u_1)],\ [(r_2,s_2),(t_2,u_2)],\ldots,\ [(r_M,s_M),(t_M,u_M)]\ ;$$

where: $w[(r_k,s_k),(t_k,u_k)] = e(t_k,u_k) - e(r_k,s_k)$; and the vertices of $[s]$ are defined recursively:

(40) $$V_1 = \bar{x} + w[(r_1,s_1),\ (t_1,s_1)]\ ;$$

$$V_k = V_{k-1} + w[(r_k,s_k),\ (t_k,u_k)]\ ;$$

and such that the following three properties are satisfied:

(41.1) For $M-1$ values of k: $(r_k,s_k) < (t_k,u_k)$, and for the other value, say k_0:

$$(t_{k_0}, u_{k_0}) < (r_{k_0}, s_{k_0}),$$

(41.2) For each k, there is some j such that:

$$(r_k, s_k) = (t_j, u_j).$$

(Hence, in the sequence (39) there are only the M distinct indices (r_k, s_k); and (t_{k_0}, u_{k_0}) is the least; (r_{k_0}, s_{k_0}) the greatest.)

(41.3) For each (n, i), $x(n, i)$ is positive on $[s]$ if and only if (n, i) is (r_k, s_k) for some k.

Hence we may note that:

(41.4) $V_M = \bar{x}$ (the base point), by summing the V_K,

(41.5) For each (n, i) such that $x(n, i)$ is 0 on $[s]$, either $(n, i) < (t_{k_0}, u_{k_0})$, or $(r_{k_0}, s_{k_0}) < (n, i)$, or, for some unique k:

$$(r_k, s_k) < (n, i) < (t_k, u_k).$$

(41.6) There is a constant column c, of order m by 1, such that, for all (n, i): for all k: $V_k(n, i)$ is $c(n, i)$ or $1 + c(n, i)$; and V_k differs from V_{k+1} (cyclically, V_m differs from V_1) in exactly two components.

One may note also that it is possible, but with some amount of shifting, to make any point the base point, in an equivalent sequence for $[s]$.

Adjacent Simplices.

We next identify, in terms of operations performed on the sequence (39) for $[s]$, the Drop and Add operations.

<u>Drop</u> V_k. Consider $V_k' = V_{k-1} + w[(r_{k+1}, s_{k+1}), (t_{k+1}, u_{k+1})]$.

<u>Case 1.</u> $V_k' \geq 0$. Then V_k' is in S ; the [s]' formed by dropping V_k and adjoining V_k' as a vertex is in the triangulation; and the desired sequence for [s]' is obtained from (39) by simply exchanging the pair:

(42) $\qquad [(r_k, s_k), (t_k, u_k)] \quad \text{and} \quad [(r_{k+1}, s_{k+1}), (t_{k+1}, u_{k+1})]$

(where, cyclically, if $k = M$, read '1' for 'M+1').

Note that then V_{k+1} , and all other V_r remain as they were.

<u>Case 2.</u> V_k' is not non-negative. Then [s]' = F (face opposite V_k) is formed by replacing the pair (42) by the single direction:

(43) $\qquad [(r_k, s_k), (t_{k+1}, u_{k+1})]$.

<u>Add Vertex</u> (increasing $x(n,i)$ from 0)

Since $x(n,i)$ is zero on [s], if k is such that $(r_k, s_k) < (n, i) < (t_k, u_k)$ (or, cyclically, if either $(n,i) < (t_{k_0}, u_{k_0})$ or $(r_{k_0}, s_{k_0}) < (n,i)$), one simply replaces the direction $[(r_k, s_k), (t_k, u_k)]$ by the two directions (in that order):

$$[(r_k, s_k), (n, i)], [(n, i), (t_k, u_k)]$$

(where $k = k_0$ in the cyclic case). The added point is thus:

$$V_k' = V_{k-1} + w[(r_k, s_k), (n, i)] ;$$

and $V_k'(n, i) = 1$.

Note that $[s]_0$ has the properties of a general [s]. From the sequence for $[s]_0$, it is easy to ascertain that $[s]_0$ is unique.

4. Remarks and Observations.

There are several observations we may make on the method, and the concepts used or developed.

First let us observe that we have visualized simplices in the bounding-faces of $S(K)$; thus of varying dimension. This, computationally, should result in some saving over always dealing with, and recording m-vertex simplices. For example, the START simplex $[s]_0$ has only N vertices.

Secondly, and relating to the above, consider case $N = 1$. The method amounts, then, to a variation of Scarf's original approximation to a fixed-point (and a proof of Brouwer's theorem, as a limiting case). In this example, given f, mapping $S(K)$ into itself, which is continuous, a label for an x in $S(K)$ is a least i such that $f_i(x) \geq x_i$. Then our method shows that there is an odd number of $\bar{1}$-completely labelled simplices. And, there is no necessity for an 'artificial start', in the sense that one would start from, $\bar{x}_0 = \langle K, 0, \ldots, 0\rangle$ and $[s]_0$ consists simply of that single point! And, again, if x is in some n-space, generally one is not dealing with n-vertex simplices. See Kuhn [5], also.)

Thirdly, generally let us observe that, implicit in the development is the fact that there is an odd number of states $(N; [s])$ where $[s]$ is N-complementary (thus, an odd number of $[s]$ which are N-complementary). Thus, from 2.1 to 2.4, since $[s]_0$ is unique, if one were to start from some possible state, either of the two permissible state changes initiates a 'path' which terminates in an $[s]$ which is N-complementary if and only if excepting loops one is not on the path described in 2.5 and not 'heading back to $[s]_0$' along that path. This statement, of course, involves no notions of 'degeneracy', since it involves paths defined in the text.

Fourthly, let us observe that there is a kind of 'uniqueness' to the method, in the sense that, given the (nature of the) START, and the desire to generate an 'almost-complementary path', based upon 'almost-completely-labelled' simplices, it becomes a question of merely identifying, for any given state, the permissible changes of state.

This is a strong statement (without proof), and would be rendered void with the appearance of a variation.

Fifthly; observe that if a better approximation is desired (e.g., using larger K), the method would have to be restarted from (the new) $[s]_0$. An algorithm similar to the one here developed which was patterned on the work of Eaves as developed in Ref. [2] would resolve this issue.

Sixth, as a possible alternative algorithm, let us consider the approach of Sobel (Ref. [9]). Sobel introduces a 'dummy' (N+1)st player, which permits a START from an (N+1)-complementary point, and terminates (in principle) with an N-complementary point. It shall be reported how a 'simplicial path' method, based upon the one developed here, and upon Sobel's approach is constructed. (Observe the dependence of the method upon the particular START available!)

Finally, there is the question of possible generalization of the results here obtained, based upon the new concepts developed; essentially introducing 'another parameter', namely p, into the almost-complementary setting.

5. Some Computed Examples.

A FORTRAN program was written and tested (for $N = 3$ or 4) on an IBM 360 model 50. We always took $m_n = N$, for each n. Various values of K were tried. The results are summarized as follows:

I. Game 1 ($N = 3$; thus 9 $f(n,i)$'s; and each f has $3 \times 3 = 9$ terms. Thus, the coefficients in (4) are $m \cdot \bar{m}/3 = (3+3+3) \times 9$ in number. Duplicate data was used.

Let n be a player; and j, k (where $j < k$) denote the remaining two players. The game data may be exhibited in matrix form as:

	[1, 1]	[1, 2]	[1, 3]	[2, 1]	[2, 2]	[2, 3]	[3, 1]	[3, 2]	[3, 3]
(n, 1) :	2	3	4	2	3	3	4	1	5
(n, 2) :	1	1	4	3	4	1	6	8	2
(n, 3) :	4	7	2	4	5	5	3	6	4

The column headings have the form $[i_j, i_k]$ referring to the i_jth pure strategy for j and the i_kth pure strategy for k. Thus, for example: (using the second row)

$$f(2,2;x) = 1\cdot x(1,1)x(3,1) + 1\cdot x(1,1)x(3,2) + 4x(1,1)x(3,3) + 3x(1,2)x(3,1) + 4x(1,2)x(3,2) + x(1,2)x(3,3) + + 6x(1,3)x(3,1) + 8x(1,3)x(3,2) + 2x(1,3)x(3,3).)$$

K = 10 was used. After 7.17 seconds, and 7 iterations (i.e., 7 vertices of the mesh were encountered) the terminal simplex:

$$\bar{x}: [(1,2)(2,1)],\ [(2,1)(3,1)],\ [(3,1)(1,2)];$$

where $\bar{x} = \langle 0,2,0; 4,0,0; 4,0,0\rangle$, was found; yielding an exact solution.

(Note: the time given includes linkage editing and initialization.)

II. Game 2. (N = 3). The same matrix was used, except that the [2, 1]-headed column was changed to $\langle 4,3,2\rangle$. K = 10. After 8.45 seconds and 54 iterations, the terminal simplex obtained was:

$$\bar{x}: [(1,1)(1,2)],\ [(2,3)(3,1)],\ [(2,2)(2,3)],\ [(2,1)(2,2)]$$
$$[(3,2)(1,1)],\ [(3,1)(3,2)],\ [(1,3)(2,1)],\ [(1,2)(1,3)];$$

where: $\bar{x} = \langle 1,1,1; 2,1,1; 2,1,0\rangle$.

III. Game 2; K = 100. After 10.40 seconds, and 432 iterations, the base point of the terminal simplex was:

$$\bar{x}_T = \langle 13,\ 9,\ 11;\ 14,\ 9,\ 11;\ 32,\ 1,\ 0\rangle$$

IV. Game 2; K = 500. After 45.15 seconds, and 2205 iterations, the base point for the $[s]_T$ was:

$$\bar{x}_T = \langle 65, 48, 53; 65, 48, 53; 163, 5, 0 \rangle .$$

The computed deviation from complementarity is small, when computed for $\bar{x}_T$. (Note that equilibrium point approximations for III and IV are for the same equilibrium point.)

V. Game 3; N = 4; K = 200. (A total of $m \cdot \bar{m}/N = 1024$ pieces of data were created.) After 2 minutes, 5.69 seconds, and 3440 iterations, the base point for a $[s]_T$ was

$$\bar{x}_T = \langle 75, 0, 16, 0; 0, 0, 23, 1; 0, 0, 0, 50; 35, 0, 0, 0 \rangle .$$

The deviation from complementarity was .1633.

VI. Game 4; N = 4; K = 200. Only slightly different from Game 3. After 2 minutes, 18.17 seconds, and 3052 iterations, the base point for a terminal $[s]_T$ was:

$$\bar{x}_T = \langle 40, 0, 15, 0; 11, 0, 39, 0; 29, 0, 0, 26; 40, 0, 0, 0 \rangle .$$

In summary, it was found that the great bulk of time was spent in the computation of the multilinear functions, which of course is understandable. No attempt was made to do this efficiently. There is probably some natural efficient way of doing this; for example, by iteration on previous values. In any case, time-saving must come in this way.

It is also no doubt quite natural that the number of iterations increases combinatorially with N. In this regard, one could observe that at any point, without having a terminal simplex the deviation from complementarity could be computed (if one is willing to compute N-labels), and iterations could cease if below a pre-assigned tolerance.

REFERENCES

1. Cairns, S. S., "Introductory Topology", The Ronald Press, New York (1961).

2. Eaves, B. C., "Homotopies for Computation of Fixed Points", Math. Programming, Vol. 3, pp. 1-22, (Aug. 1972).

3. Hansen, T., and Scarf, H., "On the Applications of a Recent Combinatorial Algorithm", Cowles Foundation Discussion Paper No. 272, Yale Univ. (1969).

4. Howson, J. T., Jr., "Equilibria of Polymatrix Games", Management Science, Vol. 18, No. 5, pp. 312-318, (Jan. 1972).

5. Kuhn, H. W., "Simplicial Approximation of Fixed Points", Proc. Natl. Acad. Sci., Vol. 61, No. 4, pp. 1238-1242, (1968).

5a. Kuhn, H. W., "Approximate Search for Fixed Points", Computing Methods in Optimization Problems, Vol. 2, pp. 199-211, Academic Press (1969).

6. Lemke, C. E., and Howson, J. T., Jr., "Equilibrium Points of Bimatrix Games", SIAM Journal, Vol. 12, No. 2, (June, 1964).

7. Lemke, C. E., "Recent Results on Complementarity Problems", in: Nonlinear Programming, Ed. J. B. Rosen, O. L. Mangasarian, and K. Ritter, Academic Press, New York, pp. 349-384, (1970).

8. Rosenmüller, J., "On a Generalization of the Lemke-Howson Algorithm to Non-cooperative N-person Games", SIAM J. Appl. Math., Vol. 21, No. 1, pp. 73-79, (July 1971).

9. Sobel, M. J., "Algorithm for a Game Equilibrium Point", CORE, Univ. Catholique de Louvain, Belgium (Nov. 1970).

10. Wilson, Robert, "Computing Equilibria of N-Person Games", SIAM J. Appl. Math., Vol. 21, No. 1, pp. 80-87, (July 1971).

C. B. Garcia, C. E. Lemke
Department of Mathematics
Rensselaer Polytechnic Institute
Troy, New York 12181

and

H. Luethi
Institute for Operations Research
Eidg. Technische Hochschule
Zürich, on leave at RPI.

Research partially supported by the
National Science Foundation Grant NSF-GP-15031

On Balanced Games without Side Payments

L. S. SHAPLEY

1. Introduction

In this paper we present a new proof of a basic theorem of game theory, due to Scarf, which states that every balanced game without side payments has a nonempty core.[†] Our main tool is a new generalization of Sperner's celebrated topological lemma concerning triangulations of the simplex, which we believe will be of independent interest.[‡]

Like Scarf, we base our proof on a "path-following" algorithm decended from the Lemke-Howson procedure for finding equilibria in bimatrix games[*]. Despite this and perhaps other similarities, we believe that our proof is not only shorter than Scarf's original but more intuitive, or at least easier to follow, since it stays close to familiar ground most of the way and specializes to the game context only at the very end. At any rate, this re-proof of an old result will serve an expository purpose for readers new to the subjects of balanced sets and n-person games; we have accordingly

[†] Scarf (1967a); see also Billera (1970, 1971) .

[‡] Sperner (1928); also Knaster, Kuratowski, Mazurkiewicz (1926).

[*] Lemke and Howson (1964); see also Cohen (1967), Scarf (1967b), and Kuhn (1968, 1969). Similar techniques are now widespread in mathematical programming.

tried to make the presentation as self-contained as possible.

The section titles should be a sufficient guide to the contents. The economic example in Section 4 may be skipped without loss of continuity. Two items of special note are (1) the simple but very useful geometric characterization of balanced sets, described in Section 3, and (2) the handy notational scheme for iterated barycentric partitions of the simplex, described in the Appendix.

2. Games and Cores

Let N denote the set $\{1,\ldots,n\}$, and let $\mathcal{N}$ denote the set of all nonempty subsets of N; thus $|\mathcal{N}| = 2^n-1$. Let E^N denote the n-dimensional euclidean space with coordinates indexed by the elements of N, and for $S \in \mathcal{N}$ let E^S denote the corresponding $|S|$-dimensional subspace of E^N. A subset X of E^N will be called <u>comprehensive</u> if $\alpha \in X$ and $\beta \leq \alpha$ imply $\beta \in X$. If $X \subseteq E^N$ then $\bar{X}$ will denote the closure of X, and $\hat{X}$ will denote the "comprehensive hull" of X, i.e., the smallest comprehensive set that contains X. If $\alpha \in E^N$ and $S \in \mathcal{N}$, then α^S will denote the projection of α on E^S, i.e., the restriction of α to the coordinates indexed by the elements of S.

In this paper, a <u>game</u>† will be an ordered triple (N, F, D). Here N is as above, F is a closed subset of E^N, and D is a function from $\mathcal{N}$ to open, comprehensive nonempty, proper subsets of E^N that satisfies

(2.1) $$D(N) \subseteq \hat{F},$$

(2.2) if $\alpha \in D(S)$ and $\alpha^S = \beta^S$, then $\beta \in D(S)$, and

(2.3) $\{\alpha^S : \alpha \in \overline{D(S)} - \bigcup_{i \in S} D(\{i\})\}$ is bounded and nonempty.

Condition (2.1) will be discussed presently. Condition (2.2) states that $D(S)$ is a "cylinder," parallel to the subspace

† Cf. Aumann (1961), Scarf (1967a), Billera (1971).

E^{N-S}. The sets $D(\{i\})$, $i \in N$, are therefore open half-spaces of the form $\{\alpha : a_i < v_i\}$; it is sometimes convenient to normalize the game by setting all the $v_i = 0$ and shifting the other $D(S)$ accordingly. If this is done, then (2.3) states that the closure of each $D(S)$, intersected with the nonnegative orthant of E^S, is bounded and nonempty.

In the standard interpretation, the elements of N are players, the elements of $\mathcal{N}$ coalitions, and the elements of E^N payoff or utility vectors. The elements of F represent feasible outcomes and the elements of $D(S)$ represent outcomes that S can improve upon, in the sense that the players in S can through their coordinated actions ensure better payoffs for themselves, regardless of the actions of players outside S.

In view of this interpretation, it would be natural to specialize (2.1) to

$$\overline{D(N)} = \hat{F}, \tag{2.4}$$

and also to assume that the function D is superadditive, in the sense that

$$D(S) \cap D(T) \subseteq D(S \cup T), \text{ all } S, T \in \mathcal{N} \text{ with } S \cap T = \phi. \tag{2.5}$$

These assumptions do not figure in our work, however, and so we do not make them here. Similarly, it is often the case in applications that the sets $D(S)$ are convex. But convexity uses the structure of E^N as a real linear space, while we shall be concerned only with the ordinal and topological structure of E^N.

The core of the game (N, F, D) is defined to be the set

$$F - \bigcup_{S \in \mathcal{N}} D(S). \tag{2.6}$$

The core represents the set of feasible outcomes that cannot be improved upon by any coalition. It is a closed set, and bounded as well if (2.4) holds or if F is bounded. The core may, however, be empty. A central problem of game theory is to determine significant classes of games that have nonempty cores.[†]

The reader with an eye for such things may find Figure 1 helpful in visualizing the foregoing definitions. The sets D(S) are represented for $|S| = 1$ by the coordinate planes, for $|S| = 2$ by the truncated "quarter rounds," and for $|S| = 3$ by the spherical surface. The core, assuming (2.4) holds, is the shaded area.

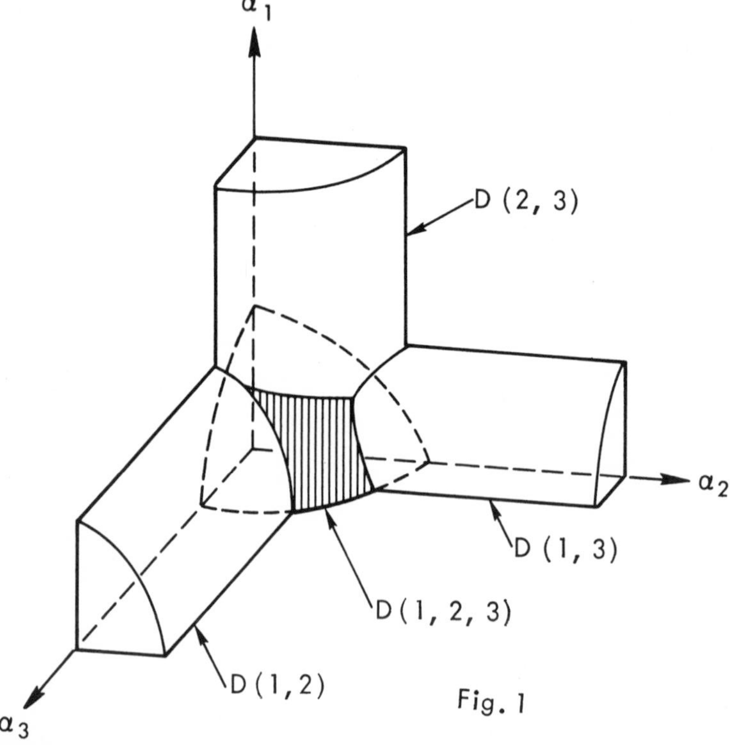

Fig. 1

[†]The games described here are "games without side payments." Games "with side payments" have a parallel but simpler theory; they correspond to games in the present sense in which each D(S) is a half-space of the form $\{\alpha : \Sigma_S \alpha_i < v(S)\}$, where v is any function from $\mathcal{N}$ to the reals. Bondareva (1962, 1963) proved (in effect) that such games have nonempty cores if and only if they are "balanced" in the sense of the next section; see also Shapley (1967).

3. Balanced Sets and Balanced Games

Let $\mathcal{B}$ be a subset of $\mathcal{N}$, and let $\mathcal{B}_i = \{S \in \mathcal{B} : i \in S\}$. The set $\mathcal{B}$ is said to be balanced (with respect to N), if there exist nonnegative "balancing weights" $\{w_S : S \in \mathcal{B}\}$ such that

$$\sum_{S \in \mathcal{B}_i} w_S = 1 , \quad \text{all } i \in N . \tag{3.1}$$

For example, $\{\{1,2\}, \{1,3\}, \{1,4\}, \{2,3,4\}\}$ is balanced with respect to $\{1, 2, 3, 4\}$ by virtue of the weights 1/3, 1/3, 1/3, 2/3. If the weights are all 1, then $\mathcal{B}$ is a partition; thus balanced sets may be regarded as generalized partitions. It is not difficult to show that positive balancing weights are unique if and only if the balanced set is minimal, i.e., has no proper subset that is balanced, and that a minimal balanced set has at most n members. Of course, any superset of a balanced set is balanced.†

Balanced sets can be given a geometric interpretation. Take any set of n linearly independent vectors in E^N, for example the unit vectors $e^1, \ldots, e^n$. For each $S \in \mathcal{N}$ define A^S to be the convex hull of the points $\{e^i : i \in S\}$ and let m_S denote their center of gravity, and hence the center of gravity of A^S as well. Then it is easily shown from the above definition that $\mathcal{B}$ is balanced if and only if m_N lies in the convex hull of $\{m_S : S \in \mathcal{B}\}$. Figure 2 illustrates.

A balanced game is defined to be a game (N, F, D) in which the relation

$$\bigcap_{S \in \mathcal{B}} D(S) \subseteq \hat{F} \tag{3.2}$$

holds for every balanced set $\mathcal{B}$. The reader can verify by inspection that the game in Figure 1 is balanced, and that if

†This is not true if positive weights are required, as in the original definition of balanced set (see Shapley (1967)). The minimal balanced sets, however, are the same under either definition.

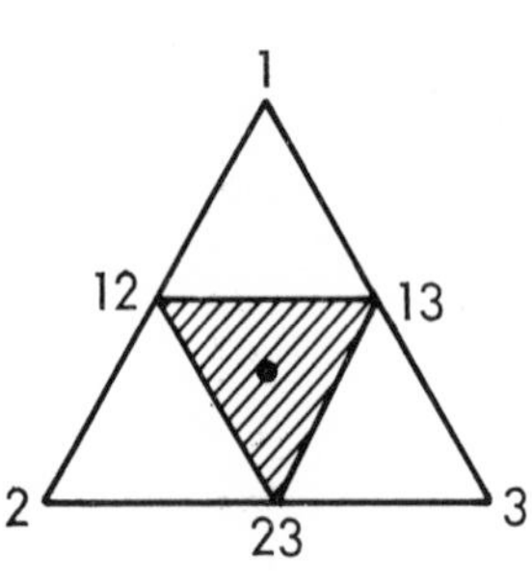

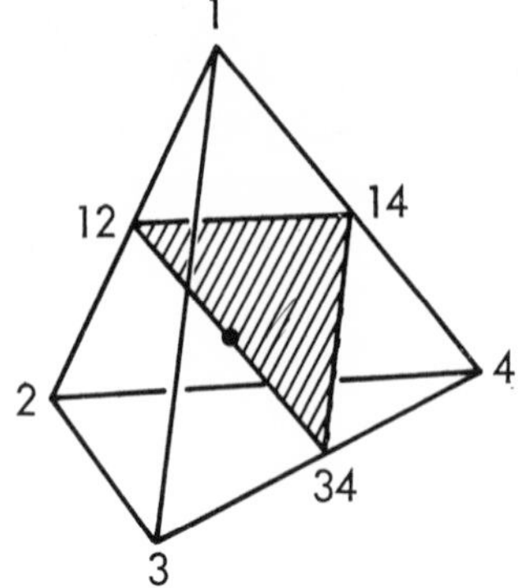

Fig. 2

the surface representing D(N) and F is pulled back toward wards the origin until the core disappears, the game is not balanced.

Theorem 3.1. (Scarf) Every balanced game has a nonempty core.

This will be proved in Section 8.

4. An Application to Economics

Balanced games arise naturally in economics, as the following model illustrates.† Let each economic agent (player) $i \in N$ have a set C^i of possible final holdings which is a nonempty compact convex subset of a linear space C. Similarly, let him have a nonempty compact convex set $Y^i \subseteq C$ of possible productions.‡ An initial holding $a^i \in C^i - Y^i$ (algebraic subtraction) is also given, and a utility function U^i from C^i to the reals, assumed continuous and

† Compare Scarf (1967a).

‡ If the reader wishes to simplify, he may eliminate production from the model by setting all Y^i equal to $\{0\}$.

quasi-concave.†

Members of a consenting group can trade freely with one another; a feasible final S-holding is defined to be a set of possible final holdings $\{x^i \in C^i : i \in S\}$ that satisfy

$$\sum_S x^i = \sum_S a^i + \sum_S y^i \tag{4.1}$$

for some S-production schedule $\{y^i \in Y^i : i \in S\}$. Thus, it is assumed that during the process each trader makes exactly one "production move," adding a selected element of Y^i to his holding; it does not matter for our purposes when this happens. Under our assumptions the set of feasible final S-holdings is convex, compact, and nonempty.

Turning to the payoff space, we define $F(S)$ to be the set of $\alpha \in E^S$ for which a feasible final S-holding $\{x^i : i \in S\}$ exists with

$$U^i(x^i) = \alpha_i , \quad \text{all } i \in S .$$

Under our assumptions $F(S)$ is nonempty and compact for each $S \in \mathcal{N}$.

The game (N, F, D) associated with this economic model can now be defined. Indeed, we merely take F to be $F(N)$ and take $D(S)$, for each $S \in \mathcal{N}$, to be the set of $\alpha \in E^N$ such that α^S is majorized (strictly) by a member of $F(S)$. Then F is closed and $D(S)$ is open, comprehensive, nonempty and proper, as required. Properties (2.1) and (2.2) can be immediately verified, as well as (2.4) and (2.5). Finally, in (2.3) the boundedness follows from the boundedness of $F(S)$ and the nonemptiness follows from superadditivity (2.5). Thus, all the defining conditions for a game are fulfilled.

Theorem 4.1. The game described is balanced.

---

†A function $f(x)$ is quasi-concave if the sets $C_z = \{x : f(x) \geq z\}$ are all convex. A concave function is quasi-concave.

Proof. Let $\mathcal{B}$ be a balanced set with weights $\{w_S : S \in \mathcal{B}\}$, and let $\alpha \in \bigcap_{\mathcal{B}} D(S)$. We wish to show that $\alpha \in \hat{F}^S$.

For each $S \in \mathcal{B}$ we can find a feasible final S-holding $\{S_{x^i} \in C^i : i \in S\}$ satisfying (4.1) for some S-production schedule $\{S_{y^i} \in Y^i : i \in S\}$ and such that

$$U^i(S_{x^i}) > \alpha_i , \quad \text{all } i \in S .$$

For each $i \in N$, define

$$x^i = \sum_{S \in \mathcal{B}_i} w_S S_{x^i} .$$

By quasi-concavity and (3.1) we have $U^i(x^i) > \alpha_i$, for each $i \in N$. Hence it remains only to show that $\{x^i : i \in N\}$ is a feasible final N-holding.

To this end, for each $i \in N$ define

$$y^i = \sum_{S \in \mathcal{B}_i} w_S S_{y^i} .$$

By (3.1) we have $x^i \in C^i$ and $y^i \in Y^i$. Finally, we have

$$\sum_{i \in N} x^i - \sum_{i \in N} y^i = \sum_{i \in N} \sum_{S \in \mathcal{B}_i} w_S (S_{x^i} - S_{y^i})$$

$$= \sum_{S \in \mathcal{B}} \sum_{i \in S} w_S (S_{x^i} - S_{y^i}) = \sum_{S \in \mathcal{B}} \sum_{i \in S} w_S a^i$$

$$= \sum_{i \in N} \sum_{S \in \mathcal{B}_i} w_S a^i = \sum_{i \in N} a^i .$$

This completes the proof.

5. Simplicial Partitions, Sperner's Lemma, and the K-K-M Theorem

A face of a closed convex set C is either C itself or the intersection of C with one of its supporting hyperplanes. A facet of C is a face of dimension one less than the dimension of C. A simplex can be characterized as the convex hull of a finite set of "affinely independent" points; a test for the affine independence of r points in E^N being that the $r \times n+1$ matrix obtained from their coordinates with a column of 1's adjoined should have rank r. A simplex has finitely many faces, all of them simplices.

As before, let A^S denote the convex hull of the unit vectors $\{e^i : i \in S\}$. Then the A^S, $S \in N$, are the faces of the $(n-1)$-dimensional simplex A^N. By a simplicial partition of A^N we shall mean a finite collection Σ of subsets of A^N, called cells, such that†

(5.1) each cell is a simplex,

(5.2) each face of a cell is a cell,

(5.3) the union of all the cells is A^N, and

(5.4) the intersection of any two cells is either empty or a face of both of them.

The mesh of Σ is the diameter of its largest cell. We take it for granted that simplicial partitions of A^N exist of arbitrarily small mesh.

Let Σ_d denote the set of members Σ of dimension $d-1$. Then Σ_n comprises the "full-dimensional" cells in Σ,

† Of course, Σ is not a true partition of A^N, because of the overlapping. But the relative interiors of the cells in a simplicial partition do form a partition.

The term "simplicial subdivision" is often employed instead, usually in reference to the subcollection of Σ we call Σ_n (see below).

so that Σ is precisely the set of all faces of members of Σ_n; we shall therefore say that Σ is generated by Σ_n. The following proposition is geometrically fairly obvious; we omit the proof.

Lemma 5.1. If Σ is a simplicial partition of A^N and if $\tau \in \Sigma_{n-1}$, then τ is a facet of either exactly one or exactly two members of Σ_n, depending on whether τ is or is not contained in the relative boundary of A^N.

Let Σ^S denote the set of elements of Σ that are contained in A^S. Then it is not hard to verify that if Σ is a simplicial partition of A^N then Σ^S is a simplicial partition of A^S; we shall say that Σ^S is induced on A^S by Σ. Moreover, if $R \subset S \subset N$ then $(\Sigma^S)^R = \Sigma^R$.

We now recall two well-known theorems. Let $V(\Sigma)$ denote the set of vertices of Σ, that is, the set of points in E^N that are extreme points of members of Σ. (Note that $\nu \in V(\Sigma)$ if and only if $\{\nu\} \in \Sigma_1$.) Let f be a "labelling" function from $V(\Sigma)$ to N, such that for every $S \in \mathcal{N}$

$$(5.5) \qquad \nu \in V(\Sigma) \cap A^S \Longrightarrow f(\nu) \in S .$$

In other words, the labels in the relative interior of A^N are unrestricted, but in the relative boundary the label on ν must be a member of the set S that defines the smallest face A^S to which ν belongs. It is convenient to define an auxiliary function F by $F(\sigma) = \{f(\nu): \nu \in \sigma\}$. If $F(\sigma) = N$ then we shall say that σ is completely labelled.

Theorem 5.2. (Sperner's Lemma) If Σ is any simplicial partition of A^N and if f satisfies (5.5), then at least one cell of Σ is completely labelled.

A straightforward limiting argument on the mesh of Σ leads to the next proposition, due to Knaster, Kuratowski and Mazurkiewicz, which can in turn be used to obtain the

Brouwer fixed-point theorem.[†]

Theorem 5.3. (K-K-M Theorem) Let $\{C_i : i \in N\}$ be a family of closed subsets of A^N such that for all $S \in \mathcal{N}$

(5.6) $$\bigcup_{i \in S} C_i \supseteq A^S .$$

Then $\bigcap_{i \in N} C_i \neq \phi$; in other words, at least one point in A^N is completely covered.

In Section 7 we shall prove generalizations of these two propositions, with the labels drawn from $\mathcal{N}$ rather than N and completeness replaced by balancedness.

6. Subbalance and π-Balance

Two extensions of the balanced set concept will be required in the sequel. The first depends on specifying a "last" element of N, say n. The set $\mathcal{B} \subseteq \mathcal{N}$ is then defined to be subbalanced (with respect to N, n) if nonnegative weights $\{w_S : S \in \mathcal{B}\}$ exist such that

(6.1) $$\sum_{S \in \mathcal{B}_i} w_S = 1 , \quad \text{for } i \in N - \{n\} ,$$

and

(6.2) $$\sum_{S \in \mathcal{B}_n} w_S < 1 .$$

This should be compared with (3.1). Note that any set of subsets of $N - \{n\}$ that is balanced w.r.t $N - \{n\}$ is trivially subbalanced w.r.t. N, n.

In our geometric interpretation, to say that $\mathcal{B}$ is

[†]See for example Burger (1963), p. 194ff, where the Sperner, K-K-M, Brouwer, and Kakutani theorems are proved elegantly in sequence. Historically, Brouwer's work (1909, 1910) preceded the K-K-M paper (1926), which preceded Sperner's paper (1928).

subbalanced means that the convex hull of the points $\{m_S : S \in \mathcal{B}\}$ has nonempty intersection with the half-open line segment $(m_N, m_{N-\{n\}}]$. Figure 3 illustrates this for the subbalanced set $\{\{1,2\}, \{1,3\}, \{2,3,4\}\}$ w.r.t. $N = \{1,2,3,4\}$, $n = 4$.

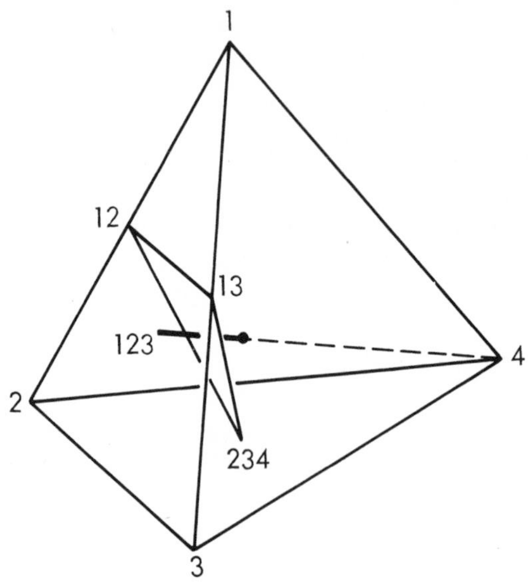

Fig. 3

For the second extension, let there be given an array of positive numbers:

$$\pi = \{\pi_{S,i} : S \in \mathcal{N},\ i \in N\} .$$

The set $\mathcal{B} \subseteq \mathcal{N}$ is defined to be π-balanced (w.r.t. N) if nonnegative weights $\{w_S : S \in \mathcal{B}\}$ exist such that

$$(6.3) \qquad \sum_{S \in \mathcal{B}_i} w_S \pi_{S,i} = 1 , \qquad \text{all } i \in N .$$

Note that because of the homogeneity of this definition a set is π-balanced if and only if it is $\bar{\pi}$-balanced, where $\bar{\pi}$ is the "normalization" of π given by

$$\bar{\pi}_{S,i} = \pi_{S,i} / \sum_{j \in S} \pi_{S,j} .$$

Ordinary balanced sets are of course $\underline{1}$-balanced, where $\underline{1}$ denotes the array consisting of all 1's.

In the geometric model, π-balancing replaces each centroid m_S by the point

$$m_S(\pi) = \sum_{i \in S} \bar{\pi}_{S,i} e^i ,$$

which can lie anywhere in the relative interior of A^S. If we let $M(\mathcal{B}, \pi)$ denote the convex hull of $\{m_S(\pi) : S \in \mathcal{B}\}$, then we see that $\mathcal{B}$ is π-balanced if and only if $M(\mathcal{B}, \pi)$ includes the point m_N. Note that "m_N" appears in this statement, rather than "$m_N(\pi)$." Thus, the set $\{N\}$, which is trivially balanced, is in general <u>not</u> π-balanced.

Combining these two extensions, we define <u>π-subbalanced</u> in the obvious way, changing = to < in (6.3) for $i = n$. The π-balanced and π-subbalanced sets will be used primarily to get around a certain degeneracy that afflicts ordinary balanced and subbalanced sets, but they will also provide us with a more general final result.

Let Π denote the set of all positive arrays π. We shall say that $\pi \in \Pi$ is <u>in general position</u> if no subset of the numbers $\pi_{S,i}$ satisfies any nontrivial algebraic equation with rational coefficients. It is clear that the arrays in general position are dense in Π, regarded as a subset of a euclidean space of suitable dimension.

<u>Lemma 6.1.</u> For each $\mathcal{B} \subseteq \mathcal{N}$, the set of $\pi \in \Pi$ such that $\mathcal{B}$ is π-balanced is closed in† Π.

<u>Proof.</u> Let $\mathcal{B}$ be $\pi^{(k)}$-balanced for $k = 1, 2, \ldots$. Suppose $\pi^{(k)} \to \pi \in \Pi^{\mathcal{N}}$ and let $\{w^{(k)}\}$ be weight vectors for the normalizations $\{\bar{\pi}^{(k)}\}$. These weight vectors lie in a bounded region in $E^{\mathcal{N}}$, so we may extract a convergent subsequence; the limit will serve as a weight vector for $\bar{\pi}$,

†Note that Π is an open cone.

showing that $\mathcal{B}$ is $\bar{\pi}$-balanced and hence π-balanced.

Q. E. D.

Corollary 6.2. For any $\pi_0 \in \Pi$ there exists a π in general position such that π-balance implies π_0-balance. In particular, there exists a π in general position such that π-balance implies balance.

Lemma 6.3. If π is in general position and if $\mathcal{B} \subseteq \mathcal{N}$ has exactly n members and is π-balanced, then every n members of $\{m_S(\pi): S \in \mathcal{B}\} \cup \{m_N\}$ are linearly independent. Moreover, if K is the affine set spanned by any n-2 members of $\{m_S(\pi): S \in \mathcal{B}\}$, then $K \cap (m_N, m_{N-\{n\}}] = \emptyset$

The first statement is proved using the fact that $\mathcal{B}$ is minimal π-balanced, the second by showing that the line including $(m_N, m_{N-\{n\}}]$ either misses K or meets it in just the point $m_{\{n\}}$. For details, see Shapley (1973).

Lemma 6.4. If π is in general position and if $\mathcal{B} \subseteq \mathcal{N}$ has exactly n members and is π-balanced, then there is a unique subset of $\mathcal{B}$ that has exactly n-1 members and is π-subbalanced.

Proof. Write M for $M(\mathcal{B}, \pi)$, the convex hull of $\{m_S(\pi) : S \in \mathcal{B}\}$. We must examine the intersection of M with the half-open segment $(m_N, m_{N-\{n\}}]$. First we note that M is full-dimensional† and hence a simplex, since otherwise the n points $\{m_S(\pi) : S \in \mathcal{B}\}$ would violate Lemma 6.3. Next we note that m_N must be interior to M, since it is in M and if it were in a facet of M then the n-1 vertices of that facet together with the point m_N would violate Lemma 6.3. Thirdly we note that $m_{N-\{n\}}$ is not interior to M, since it lies in the boundary of A^N itself

†In these proofs, "full-dimensional," "interior," etc. refer to A^N, not E^N.

while the interior of M is contained in the interior of A^N. Therefore the segment $(m_N, m_{N-\{n\}}]$ pierces the boundary of M at a unique point; call it m_0. Moreover, m_0 belongs to a unique facet F_0 of the simplex M, for if there were two such facets, then their $n-2$ common vertices, together with m_N and $m_{N-\{n\}}$, would lie in an affine set of dimension $n-2$, again in violation of Lemma 6.3. This facet F_0 determines a unique $\mathcal{B}'$ with $n-1$ members such that $M(\mathcal{B}', \pi)$ meets $(m_N, m_{N-\{n\}}]$. Q. E. D.

Lemma 6.5. If π is in general position and if $\mathcal{B} \subseteq \mathcal{N}$ has exactly n members and is π-subbalanced but not π-balanced, then there are precisely two subsets of $\mathcal{B}$ that have exactly $n-1$ members and are π-subbalanced.

Proof. The proof is similar to the previous one. The set $M = M(\mathcal{B}, \pi)$ is again a full-dimensional simplex, but m_N and $m_{N-\{n\}}$ are now both outside M. However $(m_N, m_{N-\{n\}}]$ contains at least one point of M; in fact, it contains an interior point, since a grazing contact would have to include a point common to two facets, a situation which violates Lemma 6.3. as we saw above. Therefore the segment pierces the boundary twice, intersecting a single facet each time; these two facets yield the desired $(n-1)$-member subbalanced subsets of $\mathcal{B}$. Q. E. D.

7. Generalization of Sperner's Lemma and the K-K-M Theorem

Let Σ be a simplicial partition of A^N and let f be a "labelling" function from $V(\Sigma)$ to $\mathcal{N}$, such that for every $S \in \mathcal{N}$

$$\nu \in V(\Sigma) \cap A^S \Rightarrow f(\nu) \subseteq S \tag{7.1}$$

(compare (5.1)). As before, define $F(\sigma) = \{f(\nu) : \nu \in \sigma\}$. Given f, we shall say that the cell σ is balanced if $F(\sigma)$ is balanced; similarly subbalanced, π-balanced, and π-subbalanced.

Theorem 7.1. If π is in general position and if f satisfies (7.1), then the number of π-balanced cells of Σ_n is odd.

Proof. We consider the collection $\mathcal{L}$ of all π-balanced and π-subbalanced cells of Σ. With π in general position, it follows from Lemma 6.3 that the π-balanced cells must belong to Σ_n while the π-subbalanced cells must belong to Σ_n or Σ_{n-1}. We distinguish four types of cells in $\mathcal{L}$:

a) $\sigma \in \Sigma_n$ is π-balanced. Then by Lemma 6.4 it has exactly one facet $\tau \in \Sigma_{n-1}$ that is π-subbalanced, and hence in $\mathcal{L}$.

b) $\sigma \in \Sigma_n$ is π-subbalanced but not π-balanced. Then by Lemma 6.5 it has exactly two facets $\tau, \tau' \in \Sigma_{n-1}$ that are π-subbalanced, and hence in $\mathcal{L}$.

c) $\tau \in \Sigma_{n-1}$ is π-subbalanced and intersects the interior of A^N. Then by Lemma 5.1 there are exactly two cells $\sigma, \sigma' \in \Sigma_n$ of which τ is a facet; moreover, each of these cells is π-subbalanced, and hence in $\mathcal{L}$.

d) $\tau \in \Sigma_{n-1}$ is π-subbalanced and lies in the boundary of A^N. Then by Lemma 5.1 there is exactly one $\sigma \in \Sigma_n$ of which τ is a facet; moreover, it is π-subbalanced, and hence in $\mathcal{L}$.

By the above we see that the elements of $\mathcal{L}$ are not isolated, but are "chained" together, with the "facet of" relation linking each element of $\mathcal{L} \cap \Sigma_n$ with one or two elements of $\mathcal{L} \cap \Sigma_{n-1}$ and vice versa. Each connected component of $\mathcal{L}$ therefore consists either of an endless loop, containing cases (b) and (c) only, or of a path, having case (a) or (d) at each end and all the rest (b) or (c). The important fact is that the total number of instances of (a) and (d) is even.

Let us examine case (d) more closely. Since τ is π-subbalanced, the union of its labels must include all of $N - \{n\}$. Hence, by (7.1), the only facet in the boundary of A^N that can contain τ is $A^{N-\{n\}}$, moreover, all of the labels $S \in F(\tau)$ must be subsets of $N - \{n\}$, i.e., must exclude n. This means that $F(\tau)$ is π-balanced w.r.t.

$N - \{n\}$. Conversely, any cell in $A^{N-\{n\}}$ that is π-balanced w.r.t $N - \{n\}$ is π-subbalanced w.r.t. **N**, and so falls under case (d). Hence (d) identifies precisely the π-balanced cells of the <u>induced</u> simplicial partition $\Sigma^{N-\{n\}}$ on $A^{N-\{n\}}$.

To finish the proof, let $k \leq n$ and define $K = \{1, \ldots, k\}$. Denote by a_k the number of cells in Σ^K that are π-balanced w.r.t. K. We have shown that $a_n + a_{n-1}$ is even. Similarly, since (7.1) for N implies the analogous condition for any $K \subset N$, we have that $a_k + a_{k-1}$ is even for $k = 2, \ldots, n-1$. Hence all the a_k have the same parity. But clearly $a_1 = 1$; hence a_n is odd. Q. E. D.

<u>Theorem 7.2.</u> (Generalized Sperner's Lemma) For any $\pi \in \Pi$, if f satisfies (7.1) then Σ_n has at least one π-balanced cell. In particular, Σ_n has at least one balanced cell.

<u>Proof.</u> Theorem 7.1 and Corollary 6.2.

The examples in Figure 4 show that we cannot assert that the number of balanced cells in Σ_n is odd, nor that the number of balanced cells in Σ is odd.

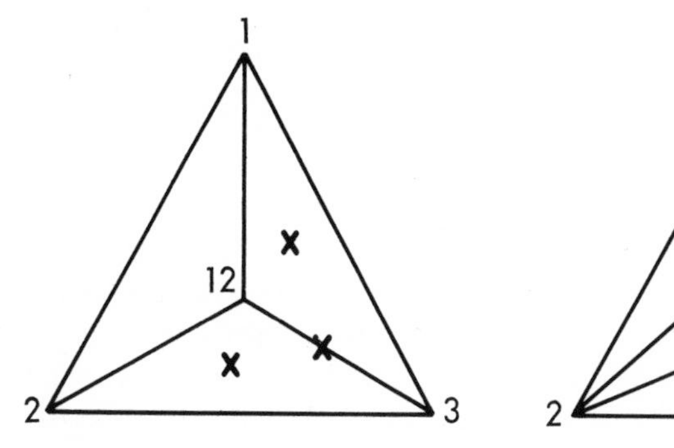

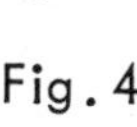
Fig. 4

To see that Theorem 7.2 includes Sperner's lemma, we restrict the values of f to the singletons in $\mathcal{N}$. Then the only balanced set (or π-balanced set for that matter) is $\{\{1\},\ldots,\{n\}\}$, and a cell is balanced if and only if it is completely labelled.

Theorem 7.3. (Generalized K-K-M Theorem) Let $\{C_S : S \in \mathcal{N}\}$ be a family of closed subsets of A^N such that for each $T \in \mathcal{N}$

$$\bigcup_{S \subseteq T} C_S \supseteq A^T \tag{7.2}$$

(compare (5.6)). Then for every $\pi \in \Pi$ there exists a π-balanced set $\mathcal{B}$ such that

$$\bigcap_{S \in \mathcal{B}} C_S \neq \phi .$$

Proof. Let $\pi \in \Pi$ be fixed and let $\{\Sigma^{(k)}\}$ be a sequence of simplicial partitions of A^N whose mesh converges to zero. For each $\nu \in V(\Sigma^{(k)})$ let $A^{T(\nu)}$ be the smallest face of A^N that contains ν and define $f^{(k)}(\nu)$ to be any S such that $\nu \in C_S$ and $S \subseteq T(\nu)$; by (7.2) such an S can always be found. By Theorem 7.2 there is a π-balanced cell $\sigma^{(k)} \in \Sigma^{(k)}$ for each k. By taking subsequences we can ensure that the $\sigma^{(k)}$ converge to some point $\nu_0 \in A^N$ and that the $F(\sigma^{(k)})$ in the subsequence are all equal to the same π-balanced set $\mathcal{B}$. Then for each $S \in \mathcal{B}$, ν_0 is the limit of a sequence of points that bear the label S and hence belong to C_S. Since the C_S are closed, we have $\nu_0 \in \bigcap_{S \in \mathcal{B}} C_S$. Q. E. D.

8. Proof of the Scarf-Billera Theorem

Although games have not yet appeared in the argument, we are close to our goal of proving that the core of a balanced game is not empty. Let (N, F, D) be a game (see Section 2), and for any $\pi \in \Pi$ let us call the game π-balanced if (3.2) holds for all π-balanced sets $\mathcal{B}$. Without loss of generality, let the game be normalized, so that $D(\{i\}) = \{\alpha \in E^N : \alpha_i < 0\}$ for all $i \in N$. Let M be a number chosen so large that for each $S \in \mathcal{N}$ and $\alpha \in E^N$

(8.1) $$\alpha \in \overline{D(S)} - \bigcup_{i \in S} D(\{i\}) \Rightarrow \alpha_i \leq M, \text{ all } i \in S ;$$

this is possible because of (2.3). Define $\gamma^i = -nMe^i$, $i \in N$; in other words,

$$\gamma^i_j = 0 \;, \quad \text{if } j \neq i \;, \text{ and } \gamma^i_i = -nM \;.$$

For each $S \in \mathcal{N}$, redefine A^S to be the convex hull of $\{\gamma^i : i \in S\}$; the new simplex A^N will provide the setting for our application of Theorem 7.3.

First we must define the sets C_S. We do this, intutively speaking, by "looking down" on $\bigcup D(S)$ from a vantage point far out in the positive orthant of E^N and observing which $D(S)$ is "on top." To make this precise, define

(8.2) $$t(\alpha) = \sup \{t : \alpha + t\underline{1} \in \bigcup_{S \in \mathcal{N}} D(S)\} ,$$

where $\underline{1}$ is the vector of all 1's. Since the $d(S)$ are proper and comprehensive the supremum in (8.2) is finite and is a continuous function of $\alpha \in E^N$. Now define

$$C_S = \{\alpha \in A^N : \alpha + t(\alpha)\underline{1} \in \overline{D(S)} \}.$$

In other words, α is in C_S if S is a "most effective" coalition along the diagonal line $L_\alpha = \{\alpha + t\underline{1}\}$, in the sense that $D(S) \cap L_\alpha \supseteq D(T) \cap L_\alpha$ for all $T \in \mathcal{N}$. Since $t(\alpha)$ is continuous, the C_S are closed sets. We shall now show that they satisfy condition (7.2).

Let $\alpha \in C_S \cap A^T$; we shall show that $S \subseteq T$. We may assume that $T \neq N$. Since $\alpha \in A^T$ we have $\Sigma_T \alpha_i = -nM$. Thisimplies that for at least one $j \in T$ we have $\alpha_j \leq -nM/|T| < -M$. Hence, considering just $S = \{j\}$ in (8.2) we obtain

(8.3) $$t(\alpha) > M \;.$$

The point $\alpha + t(\alpha)\underline{1}$ belongs to $\overline{D(S)}$ but not to any of the open sets $D(R)$, $R \in \mathcal{N}$, and in particular not to any of the $D(\{i\})$, $i \in S$. Hence, by (8.1),

$$\alpha_i + t(\alpha) \leq M , \quad \text{all } i \in S .$$

With (8.3), this yields $\alpha_i < 0$ for all $i \in S$. But $\alpha \in A^T$ implies $\alpha_i = 0$ for all $i \notin T$. Hence $S \subseteq T$, and (7.2) follows from the fact that every $\alpha \in A^N$ belongs to at least one set C_S.

Theorem 7.3 now asserts for any $\pi \in \Pi$ the existence of a point $\alpha \in A^N$ and a π-balanced set $\mathcal{B}$ such that $\alpha \in C_S$ for all $S \in \mathcal{B}$. The point $\beta = \alpha + t(\alpha)\underline{1}$ therefore belongs to $\bigcap_{\mathcal{B}} \overline{D(S)}$ but not to $\bigcup_{\eta} D(S)$. Suppose the game is π-balanced. By (3.2), β then belongs to $\hat{F}$, so there is a point $\gamma \geq \beta$ that belongs to F but not to $\bigcup D(S)$. By (2.6) γ is in the core, so the core is not empty.

We have therefore proved Theorem 3.1 in particular, and more generally[†]

Theorem 8.1. (Billera) Every π-balanced game, $\pi \in \Pi$, has a nonempty core.

9. Some Remarks on Path Following.

The proof of Theorem 7.1 may seem to be nonconstructive, but in fact it gives rise to a computationally effective algorithm.[‡] The following remarks apply equally to the problem of finding balanced cells (using Corollary 6.2), π-balanced cells, or completely labelled (Sperner) cells; we shall refer to them indiscriminately as "solutions."

Denote by $\mathcal{L}_n(a)$ the class of cells of Σ corresponding to case (a) in the proof of Theorem 7.1, i.e., the sought-for solutions, and denote by $\mathcal{L}_k(a)$ the analogous class for the induced partition Σ^K on the face A^K, $k = 1, 2, \ldots, n-1$. Similarly define $\mathcal{L}_k(b)$, $\mathcal{L}_k(c)$, $\mathcal{L}_k(d)$, and combine them all

[†] However, Billera permits some of the $\pi_{S,i}$ to be zero.

[‡] The path-following idea is implicit in the standard elementary proof of Sperner's lemma (e.g., Burger (1959, 1963)); for a very clear, explicit statement see Cohen (1967).

in

$$\mathcal{L}^* = \bigcup_{k=1}^{n} [\mathcal{L}_k(a) \cup \mathcal{L}_k(b) \cup \mathcal{L}_k(c) \cup \mathcal{L}_k(d)] .$$

In the proof we showed that $\mathcal{L}_k(d) = \mathcal{L}_{k-1}(a)$ for $k = 2, \ldots, n$. Hence each cell in $\mathcal{L}^*$ is linked (by the "facet of" relation) to exactly two other cells in $\mathcal{L}^*$, with the sole exception of the cells in $\mathcal{L}_n(a)$ and $\mathcal{L}_1(d)$. But $\mathcal{L}_1(d)$ has just the one member, $A^{\{1\}}$. Thus, if we start at that cell and simply follow the path, we <u>must</u> arrive at an element of $\mathcal{L}_n(a)$, <u>i.</u> e., a solution.† Let us call this path the <u>primary path</u> of $\mathcal{L}^*$; in general $\mathcal{L}^*$ may also contain closed loops, as well as other paths that link the remaining solutions in pairs.

Despite the dimension changes both up and down that may be encountered en route, a path-following algorithm is easy to program for a computer. It is necessary, however, to use simplicial partitions that admit a systematic description. In particular, we must be able to identify without too much trouble the cell which lies on the "other side" of a given facet of a given cell. Kuhn (1968, 1969) has described one such class of partitions; another is described in the Appendix. (See also Lemke's paper in this volume.)

The arbitrary choice of a "last" element of N in the definition of subbalance (see Section 6) gives us a chance to expand the search for solutions. Indeed, each of the $n!$ orderings of N will give us a different class $\mathcal{L}^*$ and a different primary path. Of course, if there is only one solution, all primary paths must lead to it. But conceivably we might reach $n!$ distinct solutions just by following primary paths. Moreover, whenever we find a solution that is not on the primary path for a given $\mathcal{L}^*$, we can use it as the starting

---

†Thus, primary paths provide a truly constructive proof. The proofs by induction (even Cohen's proof, though he depicts a primary path in his paper) are not constructive, since the set of <u>all</u> solutions in $A^{K-\{k\}}$ is needed to be sure of finding <u>some</u> solutions in A^K. If we are given only some solutions in $A^{K-\{k\}}$, it may happen that none of them lie on a path that leads to a solution in A^K.

point of a "secondary" path of $\mathfrak{L}^*$ and thereby reach another solution.

Were we to go deeper into the subject, we could show how to define an orientation on the solutions (including an abstract, starting-point "solution" consisting of all the $A^{\{i\}}$ lumped together), in such a way that every path in every $\mathfrak{L}^*$ has one end oriented "+" and one end oriented "-".† Thus, if we define an abstract graph G by taking the paths of the various $\mathfrak{L}^*$ as edges and the solutions as nodes, then G is a bipartite graph--i.e., it can be two-colored. If G happens to be connected, then path-following will eventually yield all solutions, if we are careful to account for all paths issuing from all solutions that we find.‡ But there is no reason for G to be connected. For example, if a balanced cell in Σ is completely enclosed by vertices bearing a single label, as in Figure 5, then there is no way for a path to penetrate the protective shell. This example shows that an exhaustive search of Σ is necessary if we wish to be sure of finding all solutions.

---

†In the Sperner case, the orientation is determined by whether the vertices of the solution cell can be mapped onto the corresponding vertices of A^N without having to turn the cell "inside out."

‡In general, many edges of G may join the same pair of nodes by the same path in Σ. Only n actual paths start from each solution, depending on which "last" element of N is chosen. Only if a path reaches the boundary of A^N does it split into n-1 continuations, depending on the "next-to-last" element of N; only if one of these hits a lower-dimensional face of A^N will it split again; etc. But no two primary paths coincide exactly, since primary paths necessarily run the whole gamut of dimensions.

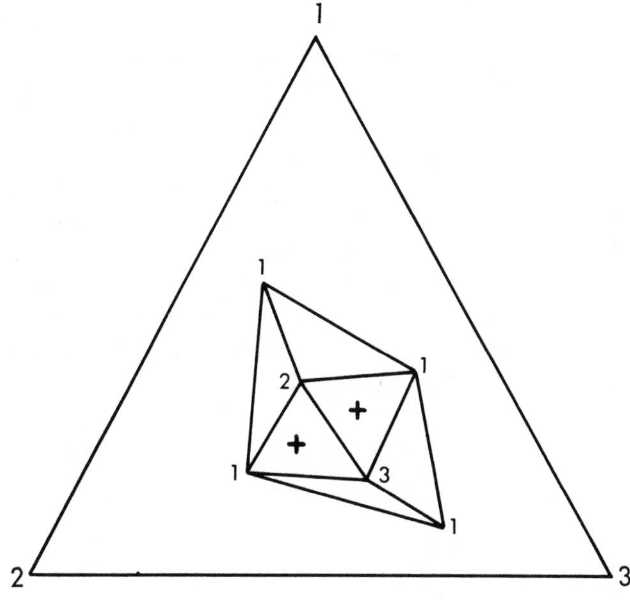

Fig. 5

Appendix. Iterated Barycentric Partitions

Let $N = \{1, \ldots, n\}$ and let A denote the simplex $\{\alpha \in E^N : \alpha \geq 0$ and $\Sigma\, \alpha_i = 1\}$. Denote by $N!$ the set of all permutations of N. If $p = p_1 p_2 \ldots p_n$ is an element of $N!$, define

$$A_p = \{\alpha \in A : \alpha_{p_1} \geq \alpha_{p_2} \geq \ldots \geq \alpha_{p_n}\} .$$

The simplices A_p, $p \in N!$, generate a simplical partition of A which we denote by $\Sigma^{(1)}$ and call the barycentric partition. In accordance with our previous usage, the collection $\{A_p : P \in N!\}$ is denoted by $\Sigma_n^{(1)}$.

The barycentric coordinates in any simplex are the relative weights (summing to 1) that must be placed at the vertices so that the center of mass will be at the desired point. In A, the barycentric coordinates of α are simply $(\alpha_1, \ldots, \alpha_n)$, because of the way we positioned A in E^N. In A_p, it may be verified that the barycentric coordinates of α are $(\beta_1, \ldots, \beta_n)$, given by

$$\text{(A.1)} \qquad \begin{cases} \beta_{p_\ell} = \ell \alpha_{p_\ell} - \ell \alpha_{p_{\ell+1}} , & \text{for } \ell = 1, \ldots, n-1 , \\ \beta_{p_n} = n\alpha_{p_n} . \end{cases}$$

The linear transformation (A.1) will be denoted by T_p; thus $\beta = T_p(\alpha)$. Its inverse T_p^{-1} is given explicitly by

$$\text{(A.2)} \qquad \alpha_{p_\ell} = \sum_{j=\ell}^{n} \beta_{p_j} / j , \quad \text{for } \ell = 1, \ldots, n .$$

We now repeat this construction. Let $p \in N!$, $q \in N!$ and define

$$A_{p,q} = \{\alpha \in A_p : \beta_{q_1} \geq \beta_{q_2} \geq \ldots \geq \beta_{q_n}\} ,$$

where $\beta = T_p(\alpha)$. If we define $T_{p,q}(\cdot) = T_q(T_q(\cdot))$, then the barycentric coordinates of α in $A_{p,q}$ are $(\gamma_1, \ldots, \gamma_n)$, where $\gamma = T_{p,q}(\alpha)$. For each $p \in N!$ the collection $\{A_{pq} : q \in N!\}$ generates the barycentric partition of A_p. Moreover, the union over all $p \in N!$ of these collections generates a simplicial partition of A, which we call the <u>barycentric partition of order 2</u> and denote by $\Sigma^{(2)}$. (Note that if A_p and $A_{p'}$ have a face in common (of any dimension), then their barycentric partitions induce the same simplicial partition on that face.)

In general, let $k > 1$ and let P represent the sequence $p^1, p^2, \ldots, p^k$. Denote $p^1, p^2, \ldots, p^{k-1}$ by P' and define

$$A_P = \{\alpha \in A_{P'} : \beta_{p^k_1} \geq \beta_{p^k_2} \geq \cdots \geq \beta_{p^k_n}\},$$

where $\beta = T_P(\alpha) \equiv T_{p^k}(T_{P'}(\alpha))$. The <u>barycentric partition of order k</u>, denoted by $\Sigma^{(k)}$, consists of all the $(n!)^k$ simplices A_P, for $P \in (N!)^k$, together with all their lower-dimensional faces.

Figure 6 illustrates this construction for $n = 3$ and various values of k. Note that each cell in $\Sigma_n^{(k)}$ receives an unambiguous name, consisting of k "n-letter words" $p^j \in N!$. Note also that the mesh of the partition decreases by at least 1/3 at each iteration. In general we have that the mesh is less than $(1 - 1/n)^k$ times the diameter of A; since n is fixed this goes to zero as $k \to \infty$.

We now number the vertices of $\Sigma^{(k)}$ in a special way. Let A_P be an element of $\Sigma_n^{(k)}$. The <u>i-th vertex</u> of A_P is defined to be the unique point in A_P whose i-th barycentric coordinate in A_P is 1, in other words, the point α such that $T_P(\alpha) = e^i$. This numbering is illustrated in the cell 123 in Figure 6. Note that the same vertex may receive different numbers in different cells; for example, we find that the second vertex of 123 is the first vertex of 213. As an exercise, the reader may verify that the point "X" is the first vertex of 231 123 and the third vertex of 213 321, while the point "Y" is second in each of these cells.

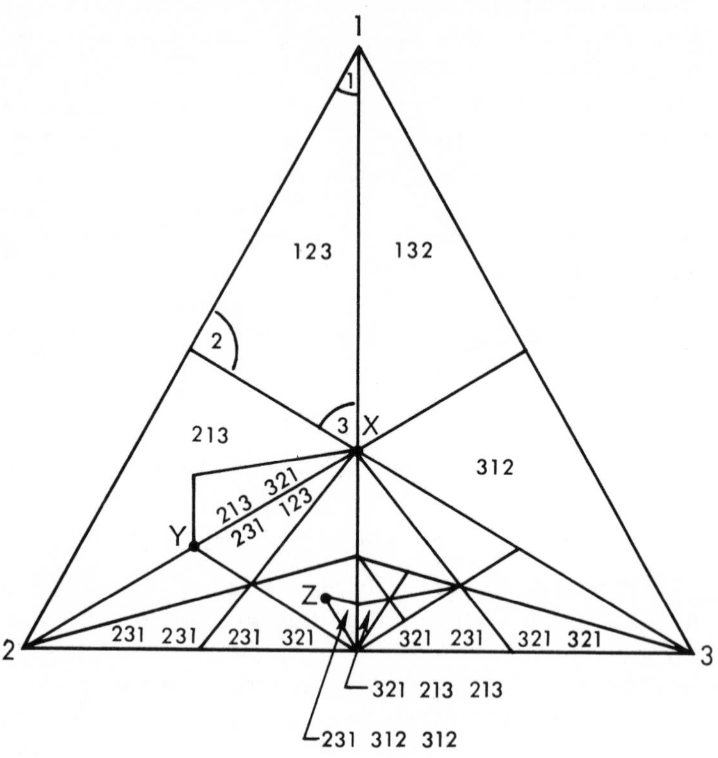
1
123
132
2
3
X
213
312
213 321
231 123
Y
Z
2
231 231
231 321
321 231
321 321
3
321 213 213
231 312 312

Fig. 6

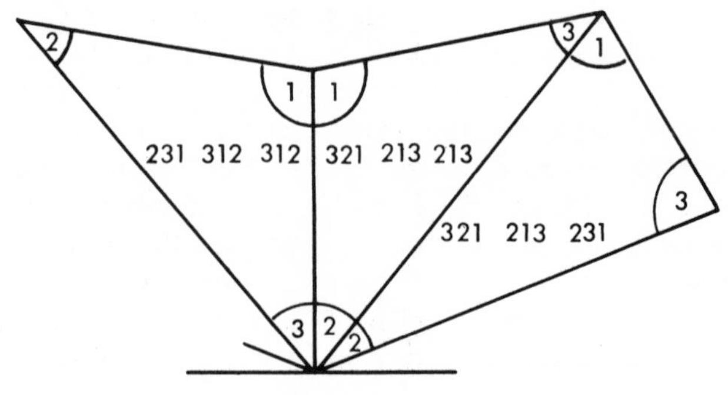
2
1
1
3
1
231 312 312
321 213 213
3
321 213 231
3
2
2

Fig. 7

The i-th facet of A_P is defined to be the facet opposite the i-th vertex, that is, the set of points in A_P whose i-th barycentric coordinate in A_P is zero. In path-following, we are interested in what lies on the "other side" of a given facet of a given cell. The rule is in fact quite simple:

FACET RULE: Let $P = p^1, p^2, \ldots, p^k \in (N!)^k$, let A_P be the corresponding cell of $\Sigma_n^{(k)}$, and let $F_i(A_P) \in \Sigma_{n-1}^{(k)}$ be the i-th facet of A_P.

Case 1: Not every word in the name of A_P ends in i. Define ℓ to be the highest index such that $p_n^\ell \neq i$ and define i' to be the immediate successor of i in p^ℓ. Then $F_i(A_P) = F_{i'}(A_Q)$, where Q is obtained from P by transposing i and i' in p^ℓ and in all subsequent words (if any). Moreover, the j-th vertex of A_Q is the j-th vertex of A_P for all j except i and i'; the i-th vertex of A_Q is the i'-th vertex of A_P; and the i'-th vertex of A_Q is the new one.

Case 2: Every word in the name of A_P ends in i. Then $F_i(A_P)$ is in the boundary of A and is not a facet of any other cell in $\Sigma_n^{(k)}$. Instead, we have $F_i(A_P) = B_{P'}$, where $B = A^{N-\{i\}} = A \cap E^{N-\{i\}}$ and the words in P' (which is a k-tuple of permutations of $N - \{i\}$) are obtained from those in P by dropping the i at the end. Moreover, each vertex of $B_{P'}$ has the same number in $B_{P'}$ as in A_P.

This rule is illustrated at several places in Figure 6. For example, in cell 231 312 312 we might want to "pivot on Z", i.e., eliminate that vertex and pass through the opposite facet to the cell beyond. Since Z is the second vertex, Case 1 applies with $i = 2$, $\ell = 1$, and $i' = 3$; the new cell is therefore 321 213 213 and its vertices are numbered as shown in Figure 7. If next we pivot on 1, we have $\ell = 3$ and $i' = 3$, making the new name 321 213 231. The reader may like to verify that three more pivots on 1 will bring the path to the boundary of A, specifically to cell 32 23 23 in the induced partition on $B = A^{\{2,3\}}$.

In a computer program, one would calculate the actual coordinates of a vertex ν only when needed to determine $f(\nu)$, using (A. 2) k times. At any given time, never more than n f-values are kept in storage, indexed by their vertex-numbers in the current cell. The dimension changes, both up and down, are easy to effect if we adopt the device of always using n-letter words, filling out the shorter words with the idle "letters" in order. Thus, we can write 23145 21345 (3) instead of 231 213, the (3) indicating that the current cell is in the face $A^{\{1,2,3\}}$ and only the first three "letters" are to be read. To "step up" one dimension we merely change the "(3)" to "(4)" and calculate the f-value for the new vertex, which will be the fourth vertex of 2314 2143. "Stepping down" (Case 2 above) is even easier since there is no new vertex to consider.

A possible drawback to the iterated barycentric partitions--as compared, say, to those used in Kuhn (1968, 1969) --is their rough texture. Most of the cells are far from equilateral (though their volumes are equal), so an unnecessary number of cells may be required to achieve a given mesh. Presumably this means that more pivot steps are needed to reach a solution of prescribed accuracy.

A possible advantage to the iterated barycentric partitions is the ease with which the geometry can be distorted to increase the cell density in the vicinity of a desired "target" point in A. Indeed, by a projective transformation of the original coordinates we can put the center node of the first partition directly on the target. Then by suitable adjustments to the transformation (A. 1), we can bring the center nodes of the second-order partitions as close as we please to the target, and so on. This geometric distortion (note that the combinatorial structure of the partition is unchanged !) would be worth the trouble if we had prior knowledge of the probable location of a solution. Such knowledge might arise from a "first pass" at the problem with a coarse grid, or from a known solution of a similar problem with slightly varied parameters (as when one is following a solution through time), or from special properties of the problem itself.

REFERENCES

1. Aumann, R. J. (1961), "The core of a cooperative game without side payments," Trans. Amer. Math. Soc. 98, 539-552.

2. Billera, L. J. (1970), "Some theorems on the core of an n-person game without side payments," SIAM J. Appl. Math. 18, 567-579.

3. Billera, L. J. (1971), "Some recent results in n-person game theory," Mathematical Programming 1, 58-67.

4. Bondareva, O. N. (1962), "Theory of the core in the n-person game" (Russian), Vestnik L. G. U. (Leningrad State University) 13, 141-142.

5. Bondareva, O. N. (1963), "Some applications of linear programming methods to the theory of cooperative games," Problemy Kibernetiki 10, 119-139.

6. Brouwer, L. E. J. (1909), "On continuous vector distributions on surfaces," Amsterdam Proc. 11.

7. Brouwer, L. E. J. (1910), Amsterdam Proc. 12, 13.

8. Burger, E. (1963), Introduction to the Theory of Games (trans. J. E. Freund), Prentice-Hall, Englewood Ciffs, New Jersey. (Original German edition published by Walter de Gruyter, Berlin, 1959.)

9. Cohen, D. I. A. (1967), "On the Sperner Lemma," J. Combinatorial Theory 2, 585-587.

10. Knaster, B., C. Kuratowski, and S. Mazurkiewicz (1926), "Ein Beweis des Fixpunktsatzes für n-dimensionale Simplexe," Fundamenta Mathematica 14.

11. Kuhn, H. W. (1968), "Simplicial approximation of fixed points," Proc. Nat. Acad. Sci. 61, 1238-1242.

12. Kuhn, H. W. (1969), "Approximate search for fixed points," in Computing Methods in Optimization, 2, Academic Press, New York.

13. Lemke, C. E., and J. T. Howson, Jr. (1964), "Equilibrium points of bimatrix games," SIAM J. 12, 413-423.

14. Scarf, H. E. (1967a), "The core of an N person game," Econometrica 35, 50-69.

15. Scarf, H. E. (1967b), "The approximation of fixed points of a continuous mapping," SIAM J. Appl. Math. 15, 1328-1343.

16. Shapley, L. S. (1967), "On balanced sets and cores," Nav. Res. Log. Quart. 14, 453-460 .

17. Sperner, E. (1928), "Neuer Beweis für die Invarianz der Dimensionszahl und des Gebietes," Abh. Math. Sem. Univ. Hamburg 6.

The Rand Corporation
Santa Monica, California 90406

Supported by the National Science Foundation, Grant GS-31253.

Index

302562767